On-Target & On-Time:
Precision Guidance for the
Military-to-Civilian Transition

On-Target & On-Time

Precision Guidance for the Military-to-Civilian Transition

Insider Secrets to Winning the Job You Want and Building a Satisfying Career

Mike Jennings, CPC

Published by Game Changer Publishing

Paperback ISBN: 978-1-966659-74-7

Hardcover ISBN: 978-1-966659-75-4

Digital ISBN: 978-1-966659-76-1

GC GAME CHANGER
PUBLISHING
www.GameChangerPublishing.com

On-Target & On-Time: Precision Guidance for the Military-to-Civilian Transition

*Insider Secrets to Winning the Job You Want
and Building a Satisfying Career*

Mike Jennings, CPC

Acknowledgments

I would like to take a moment to thank our sponsors for their financial support for this book. Many of them have been clients of Corporate Gray for many years, and I am personally grateful for their continued confidence. Please take a few moments to review the pages ahead and consider the many opportunities they have.

More personally, I would like to thank Carl and Susan Savino for their commitment to the mission of connecting employers with military veterans over the past 30 years as the previous owners of Corporate Gray. Without them, this book would have never been written this year or maybe even at all. They are living proof that you can never do a bad deal with good people, and I appreciate their above-and-beyond efforts in passing the torch to me and my team.

Most importantly, I thank God for the mission he has placed before me to live in faith in His providence while teaching others to do the same and for inspiring the work I do every day. I have always understood the importance of my salvation, but it took much longer for me to realize how much richer my life could become by aligning it with His purposes.

I would also like to thank Game Changer Publishing for reaching out at exactly the right time (a testimony of faith in providence all by itself) and for the assistance and guidance in getting this book written and published.

Many thanks to my team at Absolutely American and Corporate Gray for your focus and commitment to making sure we met our goals while we went through a great period of change and growth. Vivien, Jennifer, Scott, Anna, and Yvette: You are fierce warriors.

Finally, I am grateful for my wife, Michele, and the life we have created together on this journey. Learning how to bring two people together as one remains not only a great mystery but a constant effort, and anything worth doing will be hard at times. And I am a better man for it all.

LOCKHEED MARTIN
A career that's built to serve.
LockheedMartin.com/Military

MEET YOUR
FUTURE EMPLOYER
FIND YOUR CALLING.

Attend the Corporate Gray Virtual Military-Friendly Job Fair Events hosted throughout 2025!

Corporate Gray
POWERED BY
ABSOLUTELYAMERICAN

March 19 - April 23 - May 21 - June 25 - July 23 - August 20 - September 24 - October 22 - November 19 - December 16

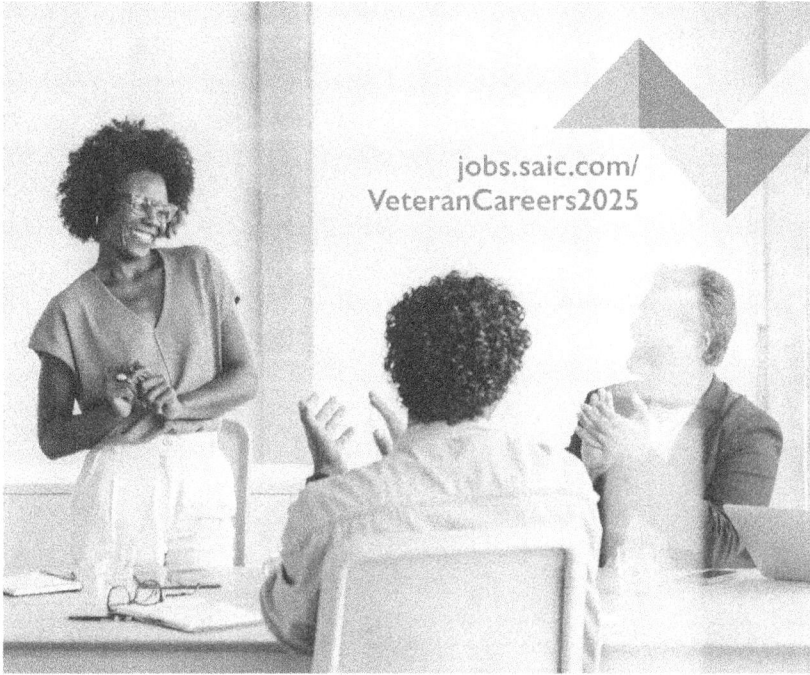

jobs.saic.com/
VeteranCareers2025

A NEW CALL
TO SERVE

Bring your valuable military experiences, skills, determination, and discipline to our team. We are a 2024 HIRE Vets Medallion award recipient. Scan QR code to learn more.

It's what happens when you **bring on tomorrow.**

SAIC.

AMERICAN
AUTOMATOR
A division of Absolutely American Inc.

Our Expertise in Technical Veteran Recruitment

As niche automation recruiters for veterans, Absolutely American has a profound understanding of how military technical skills translate to civilian roles in robotics and automation. We are adept at identifying roles that not only utilize these skills but also offer pathways for veterans to grow and excel in these innovative fields.

UNITED STATES POSTAL SERVICE®

NOW ★ HIRING
NATIONWIDE

Current Opportunities Include:
· City Carrier Assistants
· Rural Carrier Associates
· Mail Handlers
· PSE Window Clerks
· Tractor Trailer Operators
· Professional Positions

USPS® Benefits Include:
· Competitive Pay
· Diverse Workplace
· Sense of Service to the Community
· Opportunities for Advancement

APPLY TODAY
usps.com/careers

**YOU SERVED WITH THE BEST.
NOW WORK WITH THE BEST.**

Military service builds a strong work ethic and discipline that can carry into a successful career with a Caterpillar Dealership. Our customers are developing the world, but they wouldn't be able to without our equipment, engines, and support. Join our team, and continue to build your skills and the world.

Carter CAT

Contents

Special Note from the Author

It would be impossible to discuss the military-to-civilian transition without first appreciating the rights enshrined in the United States Constitution that our service members help to protect. Because of them, we get to live in a society where people are created equal, with no limits but our own. Though it has not always been that way in America, it certainly is now. Though every one of us has different starting points, none of us are limited by what we can achieve, and freedom from totalitarian governments or caste systems has provided Americans with opportunities and prosperity that remain the envy of the world.

The blessings of our liberty never became more personal to me than when my youngest brother accepted a position in Warsaw, Poland, in 2003. Growing up during the end of the Cold War, we had heard a lot about what had happened behind the Soviet Union's Iron Curtain, including stories about the government opening mail and packages and taking whatever they wanted, censoring any speech even remotely critical or threatening to the government, seizing ordinary citizens for dark room interrogations over minimal suspicions, and much worse. Prior to the dissolution of the Soviet Union in 1991, little reliable information from inside the country was available to the American public, and a part of every person thought, "This can't really all be true, can it?" Unfortunately, history has shown that all of it *was* true—and some of it was even worse than we imagined.

More personally, shortly after 9/11, my youngest brother was hired by the State Department and served at the U.S. embassy in Warsaw. As a family, we joked with him, saying, "You'd better not come home with a Polish bride!" Of course, like most little brothers, he did exactly what we told him not to do. But now, we couldn't be more happy that he did.

One of the things that we discovered fairly early on is that the nightmares that we'd heard of living under communism were truly real. My new sister-in-law told us a story about how, as a young teenager, her father told her not to attend a Catholic mass in Warsaw to avoid any notice by the Communist Party. Her father served as a colonel in the Polish Army and was among the very few who were not Communist Party members. One day, she decided to rebel against her father's wishes and attended Mass in Warsaw anyway, and she soon found out just how dangerous the Communist Party was.

The next day, her father was taken in the middle of the night. For three days, my sister-in-law and her mother had no idea what had happened to him. When he returned, he told them that he had been held captive and intensely interrogated by government officials before it was finally decided that he would be allowed to return home. The lesson was clear: Don't test the Communist Party. They can strip you of your liberty and send you to a far-off gulag without a second thought, and nobody will be able to do anything about it. You could see the fear in her eyes as she retold the story and feel the guilt she carried for jeopardizing her father the way she had. To this day, she still despises Russians. Could any of us blame her?

We live in a time that historians call the *Pax Americana*, or "American Peace"—a period where Western civilization has lived in general peace and security for more than 80 years as a result of American power and influence projected around the world. Generations have now lived and died without the imminent threat that America might be conquered by another nation that would strip us of our liberties and oppress us with real, personal fear. These were not microaggressions or harsh words. This was real oppression, where risks of physical harm, loss of liberty, and even death were ever-present, and it provided a very stark reminder of just how important our freedoms and Constitutional protections are to us. It is true to say that, sometimes, we simply don't know how important something is until it's about to be taken away.

Of course, my sister-in-law experienced a more light-hearted appreciation for America and our freedoms that night. Having grown up behind the Iron Curtain, in an economy controlled by the government of another nation, she was amazed by the variety and volume of product choices we had. Even 15 years after the Berlin Wall came down, staples as simple as peanut butter had not yet made it to Warsaw. On the deck of my home on that warm spring evening in Pennsylvania, my soon-to-be sister-in-law discovered the delight of Jif peanut butter for the very first time, and it immediately became one of her favorite things. In 2004, on her first visit to the United States, something that we treat as though it is as common as the sunrise changed this woman's life.

Recently, I heard another story from one of our contacts at FEMA. He told us how he had grown up in Cuba under Fidel Castro and how, at age 16, he had been forced into military service to fight in Angola in 1987. As a fourth-class midshipman enduring plebe summer at the Naval Academy that same year, I remember reading the newspaper stories about that conflict and how Cuban boys as young as 12 were being taken from their families to fight against an African nation that few even knew existed. At the time, I remember thinking what a terrible waste of life that was for Cuba to be doing that and how awful it must have been for those young men. I was 17 and had joined the military by choice, but they were much younger when taken from their homes to fight for something that meant nothing to their homeland.

Special Note from the Author

These are just a couple of stories that are personal to me, and they remind me of how important freedom is to our lives. When the government does what it wants to do with people, it creates a culture of fear that is simply terrible. When people live in fear, creativity dies. Innovation doesn't matter; it's only survival, except for the few who sit at the seat of power.

Freedom is the reason we can have the choices and options we enjoy in America. To know that we can live as we want, speak our minds without fear of the government taking us away, and move around the country—and even the world—without restriction is a blessing that few in history have enjoyed.

Without our armed forces and the proud Americans who serve in them, we could not enjoy these and other freedoms, and many people don't understand exactly why that is beyond defense from outside threats. You see, the officers of the U.S. Armed Forces do not swear allegiance to any leader, any party, or even to "democracy." They swear allegiance to the Constitution of the United States alone, and that small but significant distinction is what prevents the U.S. Armed Forces from being used against the population, regardless of who is commander-in-chief.

Though we enjoy freedom as Americans, the voluntary choice to serve in the military requires you to relinquish many of the freedoms you enjoy as a regular citizen and submit yourself instead to the Uniform Code of Military Justice, a much stricter standard of living.

As my chief used to tell me all the time, "We're in the business of protecting democracy, not practicing it." Military members give up much of their American freedoms as part of the service they perform, living under a certain set of laws that the rest of the people in America don't have to follow. If you don't show up for work exactly where the service wants you to, you can be put in jail—even if it means you could be killed or maimed. It doesn't matter if you're afraid or triggered. The average civilian does not understand this concept. What's more, our enlisted veterans are responsible for obeying the lawful orders of their superiors. Violate those, and they

Special Note from the Author

may find themselves locked in the brig and eating nothing but bread and water for three days.

These are the expectations that our veterans must accept as part of their willingness to serve, but it is necessary to maintain good order and discipline among those who have the responsibility to carry arms in the cause of protecting our nation and our precious liberty. Our veterans deserve respect and honor for doing the job and accepting the lifestyle that few are willing to assume—even when their freedom is on the line.

So, it is with this heart that I thank you, my fellow veterans. In writing this book, I hope to help make your return to civilian life a little easier, more successful, and a lot more enjoyable as you take advantage of the rights you have served for years to protect.

Introduction

It all started with a letter. I had had enough.

I graduated from the U.S. Naval Academy on May 29, 1991, and married the love of my life in my home church in Altoona, Pennsylvania, just 10 days later. A "stash job" at the Academy for two months provided a nice break before we packed up all of our belongings in the moving van. Then, with me driving the van and towing my car and my wife driving hers, we headed off to Surface Warfare Officers School in Newport, Rhode Island, on August 19.

In those days, weather news wasn't as ubiquitous as it is today, and we came to find out that we were following Hurricane Bob, a Category 3 storm that came ashore in Newport only six hours before we got there. As we were passing the University of Rhode Island on our way to Newport, we encountered trees down all over the roads, and when we arrived in town, all the power was out except for one hotel right across the street from the sewage plant. The night manager couldn't believe that someone was trying to check in during the middle of that night or that we had just driven nearly four hundred miles into the worst storm to hit Rhode Island in decades.

But we had no idea, and after all, orders were orders. I had to report on the 23rd, and we still needed to find a place to stay. And we did.

My orders had me set up for six months of school and then off to my ship in Norfolk, Virginia. In those days, all surface warfare officers (SWOs) went to the Division Officer's Course, with Academy guys like me joining their ROTC and OCS counterparts for training prior to going to the fleet. When we married, my wife still had a year left to finish her degree and took a year off to move with me to Rhode Island. She had made arrangements to complete her studies at a college in Southeastern Virginia, and everything appeared to be moving forward in my life as a Navy officer, just as I had planned. But that was all about to change.

Toward the end of my training in Newport, I received notice that my ship would be changing home port from Norfolk, Virginia, to the tiny town of Ingleside, Texas, about 15 miles from Corpus Christi and 1,600 miles from Norfolk. In the first of what would prove to be many Navy surprises, this one led to 14 months of living as a "geographic bachelor," as my wife returned to Pennsylvania to finish her degree because the only option in Texas was a four-hour drive away. Add in four more overseas deployments over the next three and a half years, and I was more than ready to go to shore duty.

However, during the mid-1990s, the Navy was realigning itself and decommissioning ships at a rapid rate, which is part of what led to my carousel of duty assignments. In 1994, my ship decommissioned near the end of a fiscal year when the same thing happened to more than a hundred ships, leaving me no choice on my next assignment. After being sent back to Norfolk for a one-year tour on another ship, my attempt to move to shore duty in 1995 was thwarted by losses of three sets of orders due to shore unit restructuring. With no attractive options left in my area, I volunteered to take one more sea job that would purportedly require one more deployment—a price I was willing to pay to stay in the house we had just purchased and keep my wife in her new job, which she really liked.

Well, of course, the one-deployment plan quickly turned into two in another Navy surprise, and by the time that tour was all said and done, I had, admittedly, become a little bitter and more than ready to leave the service. But quite honestly, fear kept me from trying to leave while I was still on sea duty. Add to that the fact that my sea tour would have forced me to get out immediately upon return from a six-month deployment to the Mediterranean Sea with no opportunity to interview before I got out.

Even with the help of junior officer recruiting firms, I had seen the difficulty that other officers had in trying to find a good opportunity around sea duty schedules, and I still didn't know what I wanted to do "when I grew up." At this point, I found myself at 27 years old, having never written a resume and knowing little about how to find a job I would like after the Navy—a daunting challenge you may find relatable.

I'd heard that there were a lot of great things that officers could do on the outside, but at the end of the day, when the first and the 15th of the month stopped, I didn't know what I was going to do next. Thankfully, shore duty would give me a chance to get ready, and an operational test analyst billet with the Navy's Operational Test and Evaluation Force in Norfolk, Virginia, would provide just the opportunity I needed to figure things out.

Commander, Operational Test and Evaluation Force (COMOPTEVFOR in Navy speak) is the organization responsible for evaluating the effectiveness and suitability of systems in the acquisition process that are being released for implementation in the fleet. As an operational test analyst, my role was to determine whether the C4I systems we were evaluating were properly reliable, maintainable, and available for fleet use. The nature of the position seemed to demand expertise in data analysis, but there was no Navy training for the role. I'm very thankful for my department head, Commander Mike Bond (callsign "Bondo"), who had the foresight to see that the analysts should attend some type of technical training for this particular role.

Mike had attended the Naval Postgraduate School in Monterey, California, for the Operations Research and Systems Analysis program, and while at COMOPTEVFOR, he discovered that the Army had a condensed version of that program that was taught at what was then Fort Lee in Petersburg, Virginia—an hour's drive from my house in Newport News. I was sent along with one of my colleagues to the Army's school for training post-company command Army officers transitioning into the Operations Research and Systems Analysis MOS, a 14-week, full-time, in-house course in masters-level applied math. With eight hours of class every day and a full slate of homework every night and weekend, suffice it to say that there was no time for anything but school. But I really enjoyed the challenge. Being the only Navy officers to attend this course in more than a decade also provided plenty of competitive motivation.

Though I enjoyed the program, the course on decision theory proved to be one of the most impactful on me. The class teaches students how to make logical and calculated decisions using different types of mathematical modeling in situations where an exact (or discreet) answer cannot be determined. One of the modeling techniques used "limit lines," where the answer was definitely on one side of a particular line but not on the other. This left an area of the chart where a number of "right answers" could be found, and that was the "Eureka!" moment for me. Trying to figure out *exactly and precisely* what I wanted when I didn't even really know what was out there for me was a problem that was far too difficult to solve.

I also realized that the process of figuring out what I wanted in life was not going to lead to a very specific, discreet answer. There were going to be a number of things that would fit me well that would all be closely related. Changing my process from the impossible task of trying to figure out what I wanted from everything that could be available to me and replacing it with a process that simply eliminated everything I didn't want to do was liberating, and it made the process much easier to execute. What I wanted to do came down to figuring out everything I did *not* want to do. Whatever was left would ultimately be the right answer for me. Interestingly enough, my heart had already been taking me in the right direction, but my head was telling me something different.

Coming out of the Army class, I worked hard to develop my plan to figure out what I wanted in my return to civilian life by eliminating everything I did not want. Doing my best to delay the inevitable, I submitted a letter resigning my commission to the secretary of the Navy at the latest possible moment and the longest possible timeline (one year) to avoid being sent back to a fleet unit, but this wasn't the letter of significance I mentioned at the beginning of this section. I had decided that I was getting out long before I came to shore duty, but I was still working through the process of what I wanted next. In the meantime, I had been maintaining contact with the junior military officer (JMO) recruiting firm that had placed my roommate from my first ship five years before.

Right after I submitted my letter, I met with a recruiter from that firm, a West Point graduate and former Army officer named Jim, when he came to Norfolk for a visit. During the course of that conversation, I informed him that I had just told his company that I had submitted my letter of resignation and was expecting to be getting out the following year. We talked about my career plans, and though I was a little put off by his attitude toward my career goals and search limitations, I felt like it was a good meeting.

Within a week of meeting Jim, I got a letter from his firm saying they were opening a new office in Norfolk, and they wondered if I would be interested in doing the kind of work they did, and I was. It sounded great, but after finishing the letter, I thought, *Well, I just met with Jim. They just found out that I still have a year to go, so they're not serious about me right now. Clearly, they still needed to update my file.* So, I did nothing. I just put the letter away and held on to it because I was definitely intrigued.

Three months later, I got a second letter, and at that point, I thought, *They definitely know that I'm not available for another nine months, so if they're sending me another letter now, I am absolutely going to reach out to them.* And I did. It was this letter that made all the difference.

At this point in my transition planning, I realized that I still didn't know specifically where I was going, but I did know the general fit—things that fell into that fuzzy area on my boundary line chart showed me where my solution could be found. For me, that area centered on highly incentivized sales roles and entrepreneurship. The freedom and unlimited income that came with that life had a lot of appeal to someone who had grown up humbly as I had, and when I learned about the sales opportunity with that military recruiting firm, I was hooked.

Though there were a lot of ups and downs over the next seven years, including the severe military recruiting challenges caused by the dot-com crash of late 2000, followed by stop-losses associated with 9/11, my time with that company included some of the best days of my life, and I'm grateful to this day for the opportunity.

Ultimately, I have been both an employee and entrepreneur in sales and recruiting and have enjoyed much success and satisfaction in both. The time I spent developing the boundary lines between acceptable and unacceptable career and life options has changed my perspectives in so many positive ways. Executing that process moved me from a frustrated and sometimes bitter junior officer, bored to tears unless I was driving the ship or fighting the battle, to a professional recruiter who finds myself at the end of a 10-hour day wondering where all the time went. My entrance into this profession has been truly life-changing for me and for my family. That doesn't mean that life has been easy or that I never wake up not wanting to get started, but once I show up and get rolling, those feelings disappear like they had never been there.

When people say, "When you love what you do, you'll never work a day in your life," they are wrong, but I understand the sentiment. I love what I do, but I also work very hard at it. So, what is the difference? When the work you do gives you energy instead of draining it, working harder and longer is both enjoyable and fulfilling. It's still

Introduction

work, but doing the work makes me feel better than when I'm not working. That is what we call a "life-changing transition."

So, what keeps me doing what I'm doing? Well, I got to where I am today because of a connection I had with a movie from 1996 that many people may recognize: *Jerry Maguire*. It's one of my favorite films because it speaks to the essence of the work I do.

If you don't know this movie, it chronicles the story of a sports agent, Jerry Maguire, played by Tom Cruise, who finds a heart for people in the midst of the cutthroat rat race of sports recruiting and big-money contracts, leading him to write a life mission statement emphasizing the value of people in their process over money—the thing everyone thinks but never says. After distributing his mission statement to all his colleagues and earning their applause, Jerry is terminated by his best friend in the firm because management is convinced that he has lost the edge needed for success in their dog-eat-dog business.

Determined by his competitive nature to succeed on his own, Jerry races his friend to keep his clients but loses all but one: Rod Tidwell, a chatty wide receiver (played by Cuba Gooding Jr.) with a chip on his shoulder coming off of his rookie contract with the Arizona Cardinals. In his next contract, Rod is determined to get what he calls "the kwan," something he describes as "love, respect, community, and the dollars, too." It's a four-part answer to what he describes as the ultimate value of respect and enjoyment for the work he performs at great risk to himself.

When I watched the movie the first time, with Jerry growing to become the person he described in his mission statement and Rod rekindling his love for the game to ultimately get "the kwan" he wanted, it made me realize that leaving the military is a lot like entering free agency as a professional athlete. When you come into the service, everybody has a restricted commitment. You can only play for one team: the Army, Navy, Air Force, Marines, or Coast Guard. During that time, you are obligated to play for only one team; you can't go to work anywhere else. At the end of this obligation or succession of obligations—whether it's two years, three years, five, 10, 20 or whatever it is—you enter a period where you can work for anybody. You can stay with your current employer if you have done the right things, or you can play for any other employer out there.

That is the same thing that happens with an athlete. The key to getting the kwan, as it were, is being able to get more people to want you on your team. In the case of the movie, a personal obsession with self is what prevented Rod from getting any team's interest. It was only through redeveloping a love of the game and a focus on being a teammate that he was able to develop the interest of his current team and others and accomplish his goal.

A military career works the same way. If you'd like to have options inside and outside of the military, you have to develop an approach that leads to more people wanting you to be a part of their team. Therein lies the trick. In both sports and military free agency, there is a day when your contract ends and you stop getting paid. Approaching that day does funny things to your head. It's very challenging to go through the process of setting aside the things that are very important to you in order to focus on another person's needs enough to get them to want you to work for them. That's why I've written this book. I have personally helped veterans, officers, and enlisted alike make the transition from military to civilian careers for the past 25 years.

You may ask, "How can I be certain that what you teach works?" That is a fair question. The answer is simple. In the world where I have operated, not only do the veterans I coach succeed in getting the job they want and the best offer to go with it, but their employers pay my company thousands of dollars for our services to make that happen. So, with that knowledge, I return a question to you: "If the techniques I teach work when employers have to pay my firm a sizable fee, how much more effective will they be when you use them on your own?"

As a recruiter and career coach, I determine the current qualifications and level of talent of each person I evaluate and help that

Introduction

person develop a market for that talent that leads to a role that brings out their very best. This book is for anyone making a career transition, but it is especially for veterans transitioning out of their active-duty contracts—whether as "rookies" or after a full career—entering the "free agent" world for the first time. If it weren't for the freedoms you have served to defend, we wouldn't have this free-agent world full of opportunities just waiting for you.

In the time we spend together, I hope to give you the tools you need to make a life-changing, military-to-civilian transition. As you know, we veterans love clear, step-by-step processes and clarity. From process and clarity, we develop confidence to move forward. In many cases, I have found military members to be far more confident and willing to charge into battle than to pursue the military-to-civilian transition process. I was among this group because I had no training or understanding of the expectations.

In our combat preparation, we know what to do and what to expect because we have trained for years to execute it well, but when it comes to getting out, we find ourselves with much to do and learn in very little time. So, it is my hope that this book will help you understand the world you're entering, give you the tools you need to be successful in making your plan, and inspire confidence in your ability to launch your civilian career successfully and change your life in a deeply personal and positive way.

Chapter 1
The Break-Up

So, you've decided to get out and go back to civilian life, or maybe it's been decided for you. Either way, your relationship with active duty military service is coming to an end whether you made the call to end the relationship or the service did, and like all break-ups, this can be painful. Life goes on, and either way, the question you still have to answer is, " Now what?"

When it comes to leaving active duty, the challenge for all service-members leaving active duty remains the same: How do you find gainful employment that uses the skills you have, provides the income and lifestyle you need and want, and offers growth opportunities to keep you from stagnating?

First, understand that the military-to-civilian transition is the first step in the process of building your career, not the last—and it is most definitely a process. Certainly, it would be great if every person found the perfect job with the perfect company offering perfect growth for the remainder of their working days. Then everyone could stay in one place with one organization and never have to find another employer again. That may be the path for an extremely small number of highly specialized veterans, but it is a fantasy for nearly every other person leaving active duty.

The reality is that most of us veterans will not only have several employers and several jobs as civilians, but we will likely also have several career paths. Though that may seem a bit daunting right now, it's actually an incredible blessing. As we explained in the Introduction, the military-to-civilian transition is similar to that of a professional athlete coming off their rookie contract. As a veteran becoming what we call a "Military Free Agent™," there are a vast number of organizations that you can work for that offer unique and interesting opportunities.

At the same time, all of these organizations are going through different cycles of ups and downs as part of a normal economic life cycle. Inside these organizations are people who are trying to grow them, make them more efficient, or get them turned back in a positive direction. People having the freedom to competitively solve problems stands out as one reason why market-based solutions work. Even the government goes through cycles of hiring a little and a lot. Some government agencies can be hiring while others are not.

So many of the reasons that drive the change we seek after the service frequently have their roots in why we decided to join in the first place. Maybe you had a desire to serve our country that came from your sense of patriotism or service to our nation. Perhaps military service is a tradition in your family that you wanted to continue. Maybe other members of your family have served, and you might even have felt pressured to serve because of that. Some people join because they need direction in their lives. They feel unsure about what to do after high school, and the military appears to be a good choice for figuring that out. Other people sometimes join the military for purely economic reasons, ranging from needing a job to getting education benefits or even gaining an advantage in career success. And, of course, some join with the intention of making a full career out of it.

If you completed your schooling or hit a point in your life where jobs in your area were scarce, maybe you thought that a branch of the armed forces was one employer that could address your needs. Some people join for the education and training benefits such as the GI Bill, ROTC scholarships, service academies, training to develop a trade, and more. Of course, others are simply attracted to a military career and the options it provides. You wanted to be a pilot. You wanted to be a technician. You wanted to be an engineer on a ship. You wanted to lead troops into military operations. Whatever it might be, there's a reason you came in, and that reason is usually connected to why you decide to get out.

In most cases, there's a significant contrast between who we are when we enter the service and who we are when we leave it. Most military members come in being responsible only for themselves. Usually, there's minimal debt, if any at all, and the person is young, typically under 23. It's usually just them that they have to care for, but by the time they leave the service, most find themselves with a lot more responsibilities.

You start to gather stuff. You have a car. Maybe you have a house. Maybe you have a family. You have a dog. All these things, you didn't have when you came in, and the feeling of pressure that can come with having to address those responsibilities in your military-to-civilian transition will mess with your head. That makes this process very tricky. Balancing the pressure of the situation with the faith that everything is going to work out does not come naturally to most people.

Add to all of this that most of us will miss a number of things about our time in the service that we don't often consider. Among these is esprit de corps. If you're not familiar with that phrase, it means enjoying the company of fellow service members and liking your job more for it. Veterans categorically tend to enjoy the connection we have with other service members in our communities. After all, spending a number of years on the other side of the planet and sharing a room or tent with a hundred of your closest friends can do that.

Another thing people tend to miss in leaving the service is having a strong sense of purpose. Tying one's work to protecting our nation and some of the other missions the military performs adds importance and greater purpose to military service and, by extension, gives a person a feeling of relevance. For special forces operators, the stark contrast between the intensity and significance of their military lives and the perception of the insignificance of their civilian work has led to suicide and depression rates that far exceed those in the regular services.

Human beings need to feel significant to be healthy, and tying your job in the military to your view of your value as a person can lead you down a very dark path. After all, wearing the uniform gives people a sense of pride, and our training encourages that. The uniform causes you to stand out in a crowd, and people respect you for it. When you step away from that, it is easy to feel like you don't have as much relevance as you did when you were in uniform.

Additionally, the military provides a high degree of stability and predictability to your life. You know exactly when you're going to get paid. You know exactly what you're going to be paid. You know clearly what the standards are for being able to be promoted, and if you don't know them, there is a manual where you can find the answer. Though many leave the service feeling the need to get away from these things, their impact on the psyche of the veteran leaving the service has to be recognized.

The military is built around a series of processes, standard operating procedures, and written instructions for everything service members do, and there is comfort in that. Of course, these standards exist because the armed forces must constantly onboard and train people to be effective in a system in very large numbers across a wide variety of functional specialties. When re-entering the civilian world, expect to find much less structure in civilian employment in all but the most highly regulated industries. Many who leave the service after a full career struggle with the lack of structure and bureaucracy outside of the military or government environment.

Then there are benefits such as the commissary and exchange. Having the privilege to shop in the commissary provides an enormous cost-saving benefit, and tax-free shopping is always great. Base MWR benefits are also easy to miss when you get out, whether it's recreational facilities, the auto shop, or any of the number of benefits that base access provides. Then, of course, the untaxed portion of your pay and benefits is nearly impossible to duplicate outside of the armed forces.

Beyond that, you don't have to figure out what you're going to get paid because you just quote the pay chart and move from station to station to find what your new BAH is going to be. So, the consistency of the pay and the free health, dental, and vision care for you and your family are among the benefits that you will miss as a veteran. Despite this, many service members get out anyway. Ultimately, unless you are killed in the line of duty, you will eventually transition out of the military.

Military service is a young person's career path. Even people we refer to as "retired" after 30 years are typically 48 to 54 years old and have a lot of career life left, which makes the term "retirement" from the military a real misnomer. If you enlist right out of high school, a 20-year military career makes you "retired" at 38—hardly the end of your working life.

However, there has to be something outside of military service that is more valuable to people than what is on the inside to make so many of us veterans want to go through the process of the military-to-civilian transition. The key is replacing the fear of the unknown with the confidence that comes from learning how to make the transition successfully.

～

Due Diligence for Life: How Smart Planning Turns Risk into Opportunity

Investors and business leaders manage the risk of any venture they're pursuing by a process called "due diligence." This is just a fancy term for doing your homework to understand what you're getting into: examining risks and how to deal with them, counting the costs and figuring out how to pay them, and determining if the benefits are worthwhile in comparison. Ultimately, a good acquisition, business venture, or investment is found using a due diligence process that helps you stack the odds in your favor. When it comes time for you to execute, you don't have to be afraid of what is going to happen because you have figured out how to get many of these risks out of the way before you even start the process.

Now, the trick to having a great life is refusing to settle for what you want while being willing to follow a due diligence process to get it. Business guru Jim Collins is famous for his maxim, "Good is the enemy of great." He observed that just as good businesses are resistant to making a move to become great, people are equally resistant in their personal lives to move past the perception of a good life and into a great life. It is the nature of people that we tend not to want to change what we're doing unless there is significant pain compelling us to do so. Living a pretty good life means that pain is either absent or well below your tolerance threshold. With no compelling reason to react, the status quo reigns supreme.

But here is an irony I see every day: Though most veterans tell me that one of their greatest fears is getting locked into a dead-end job, seeking change only when acted upon by outside circumstances is the perfect prescription for landing in a dead-end job and a stagnated career and repeating bad decisions over and over. We will explain the concept and balance of "running from" and "running toward" later.

Starting your due diligence process well in advance of your transition will help make it successful. It provides the time for you to carefully consider what you really want (knowing your target) and how to get it. For some, this may be easily accomplished in the first step after active duty. For others, it may require one or more interim steps. For many, it depends on where you are when starting the process, and connecting you from where you are now to where you want to be is ultimately what the military-to-civilian transition process is all about. Taking the time to create a realistic and workable plan to get there replaces the stress that comes from fear of the unknown with the confidence that comes from developing a well-considered plan.

Getting It Together *Together*

If you are married or have a significant other in your life and are pursuing the military-to-civilian transition with someone else, I cannot overemphasize the importance of getting together and talking through the things that are important to each of you in both the short and long terms. Avoid getting into specifics on jobs and locations of interest before the two of you are first clear on what is important to you individually and collectively.

It has been my experience that coming together as a couple on value systems helps to reveal the things that are most important to you both while creating room to find the places where short-term sacrifices can be made to reach long-term goals. Opinions on the specifics of jobs, employers, travel, and places to live can change over time, but your value systems provide a solid basis for decisions about your future that you will both be able to support.

When you and your partner can get connected with your value systems, figuring out where you want to go becomes much easier. For example, I landed in sales and entrepreneurship because not having somebody else tell me what my efforts were worth or how much I could make was really important to me.

Thinking about this more deeply, I have never liked having limits imposed on me; therefore, I wanted to be able to create value above and beyond an hourly-based or a time-based pay plan. For me to live my value system, I had to become comfortable managing a certain degree of downside risk, and that required an investigation (aka due diligence) and planning to move forward in roles that compensated this way.

My wife has always valued stability in our lives, whether we have much or little, and we both value close, familial relationships and want to extend that benefit to our children. These values guided us in making choices we could both support. Short-term sacrifices proved easy to make once we were both confident that they would help us have not just a good life but a great life, and pursuing our goals over time has proven this correct.

Of course, I think we all can question our choices for a time, and I did, too. In what I now call my "sabbatical year," I tried working in a sales role that offered a generous base salary, a lot of exceptional benefits, and stability. I hated almost every part of it. It was the worst possible thing that I could do. For me and my value system, having that defined base with minimal upside earnings stripped me of all of my initiative because I gave up the control and upside potential that truly motivates me. As a core motivation is working for the upside, accepting a role that limited my earnings drained the life out of me. In spite of how good this company and manager were to me, a year of working in this field was more than I could take.

I understand that not everyone values the same things I do, and when it comes to my core motivation of only wanting to work for upside gain, experience has shown me that I share it with very few people. Most of the people I work with every day share none of these values with me, which is fine. As long as you define your value system and make decisions consistent with it, there is no wrong answer. The world has opportunities for every motivation and value system. So, the maxim "to thine own self be true" proves valid even when it comes to the military-to-civilian transition.

TAP Your First Resource

To better help service members transition into civilian life, the U.S. Department of Defense, Veterans Administration, and Department of Labor joined together to standardize and rebuild the Transition Assistance Program (TAP) in 2012, and they established three training tracks most commonly pursued by veterans leaving active duty: finding a job, pursuing full-time school, and starting a business. Though these three tracks may or may not still be available at your base, your first big decision will be choosing which you want to take. For the record, "finding a job" has always been the most attended TAP track.

Your next big decision will be about where you are willing to live. Very few military members find themselves living in their hometown as their last duty station. Therefore, answering the question, "Do I stay, or do I go someplace else?" may be the most consequential of all the planning decisions you make, and science works against you in this. The law of inertia says, "An object at rest stays at rest unless acted upon by an outside force, and similarly, an object in motion stays in motion unless acted upon by an outside force." When it comes to where you live, staying where you are is always the easiest thing to do because it can be accomplished with no thought at all, but it can also be the most damaging to your career.

If you decide to move, determining whether going to a specific location right away fits your values best or if other locations are possible is an interim step. Not surprisingly, moving back to one's hometown draws many whose value system places high value on family, but in my experience, forcing that move when leaving active duty has caused the failure of more military-to-civilian transitions than I care to remember, because the veteran allowed that value system to dominate all others without considering the implications of that decision. Frequently, this decision proves worse than remaining in the area of your last duty station, particularly when moving to small towns or economically depressed areas. Balancing the desire to return home with the rest of your value systems will better inform your decision process and help you get home the right way, which we will discuss in later chapters.

The government relocation benefit contributes to the draw to move home because the government has a responsibility to return you to where you entered the service at no cost to you, subject to limitations based on family composition and size. This law has been interpreted to mean that the benefit can be applied to any location within a distance radius from your final duty station to where you entered active duty. If you go beyond this distance, your complete benefit remains available to you but may require you to pay for transportation costs above the locations inside of the radius.

Homeownership poses another challenge when deciding to leave the area of your current duty station. Choosing to leave the area usually means deciding whether to sell your home or hold it as a rental. All of these factors need to be considered when making your plan because they impact your potential income and expenses. So, of course, you're going to consider monetary factors with this transition.

Budgeting for the Change

Budgeting can be the power behind your change because it will take the largest amount of stress out of the transition process. So, what does that look like?

The first thing to consider in your transition budget is the monthly mandatory expenses that will exist, regardless of whether you remain in your current location or relocate. Housing costs typically represent the largest single item on any person's budget. Consider your housing budget before, during, and after your transition to see if there is an opportunity for savings. If you own your home, housing costs usually remain fixed until you sell or rent it.

Unless your value system includes a commitment or high likelihood of returning to the area, selling your home as part of the transition process usually proves the best option in the long run. And unless you are already a skilled real estate investor, renting a single home at a distance will pose significant challenges even with a good property manager, as issues with the house will put a strain on your home budget, add stress to your life, and take time away from your new career.

Bottom line: if you cannot afford the monthly costs of carrying your home without a tenant on top of the living expenses at your new location or have a compelling financial reason to hold the property, selling is usually the best plan. Though I am a seasoned real estate investor, renting a single-family home at a distance proves challenging, even for me, because the actions of any single tenant can determine the success or failure of that investment. This means that risks to time, money, and state of mind can be very difficult to measure and manage.

Car payments can be another major budget item that causes problems for the military-to-civilian transition. Though the mobility that a vehicle provides is an absolute necessity for most people seeking a job following active duty, an unusually large car payment can put unnecessary pressure on your budget. If you have a car payment, consider what other acceptable car options are available and if savings can be realized.

So, before you make any decisions related to your budget, examine the impact of your decision and ask yourself, *Does that budget decision empower your transition plan and give me more options and flexibility, or does it take away options and restrict my ability to make a move?*

One of the worst decisions that I have seen military members make has been to use their VA loans to buy homes in the areas where they were stationed in the final few months of their enlistments. When I inquired as to why they made that decision, they replied that it was because they wanted to buy a house while they still had their government income on record. Though that may sound like a good strategy for actually acquiring a house, it lacks foresight in figuring out how to actually keep the house once the military income is gone.

For most people, keeping a house requires maintaining earnings at a similar level to what was being made on active duty, but restricting yourself to a location based on buying a house at the end of your time in service is the kind of budget decision that restricts you rather than empowers you. Your VA loan will still be available to you when you get out. It's a great benefit of your military service when used wisely.

For more information on sound budgeting practices, I highly recommend Dave Ramsey's Financial Peace University course, which can be found at DaveRamsey.com. His team has helped tens of thousands of people to financial success through small, simple steps and commonsense approaches to saving and spending.

What Do You Like to Do?

The next phase of planning your transition is to decide what work you want to do. Regardless of the TAP track you take or if TAP continues to offer multiple tracks, all roads lead to some type of work function. Full-time education is a work function that ultimately leads to finding a job that requires the performance of some other kind of work function. Education simply helps someone get a job, either a more advanced job than what she currently has or a new job.

Being an entrepreneur will have you doing some type of work. You will just get to control what work you do. So, using your current and past roles to figure out what functional duties you have enjoyed is one of the most important parts of this process.

Here is a simple way to really understand what type of work you want to do. Start with your current role and think about the functional duties you perform every day, week, or month. What are the things you most like doing? These are the things that give you energy or get you excited when you get the chance to do them. Which ones of these would you like to do more of if you could? Write them down. We start with the positive because we want to direct our minds toward creating things we want.

Next, write down the work functions you would like to add to your job that you have not been able to do. The things of greatest benefit to you will likely be those that are extensions of the functions you already said that you enjoy doing. These are things that are exciting to you, seem like the right next step for you, and make you feel like you are moving to the next level.

Finally, write down all the things you do that you would like to eliminate from your work. These are typically the things that you feel take energy from you. Even though you can do them, they're hard for you to motivate yourself to do over and over again. These are the things that require you to apply your military discipline to be successful with them. In other words, they don't play to your strengths or interests but shore up your weaknesses.

Repeat this for each of the jobs you have had in the military, and by the time you are done, you will have a great picture of the most important things that need to be a part of your work for you to be successful, fulfilled, and motivated.

Periodically using this three-step process to look at your job will help you direct your career. As you constantly work to do more of what you want and add things that excite you, you will consciously eliminate the things you don't want to do and find your way to happiness and fulfillment. You will no longer be waiting for something to act on you to find the dissatisfaction necessary to make a change. Instead, you will be consciously choosing great over good while you eliminate the bad. This helps you create dissatisfaction with the status quo and will prompt you to make changes. The status quo you want to keep is the continuous process of intentionally reviewing where you are with a mindset of always trying to make your life better. Why is this so important? Because we spend so much of our lives at work.

The Time We Spend On Work

The standard work life of an average American is 40 hours a week of actually doing the job. On top of that 40 hours, however, between 30 minutes and an hour of each day is set aside for lunch. And while you're on a lunch break, you're still thinking about your work to some degree. So, that means 8 1/2 hours a day, at a minimum, are tied to thinking about your work.

Then there's the commute. I've read recently that the average commute for an American is 27 minutes each way. If we round that off to a half-hour, that's another hour per day that we spend going to and coming home from work. Then there's the time you spend getting ready for work each day. Typically, the moment you wake up in the morning, your entire purpose is getting ready to go to work at that particular time. That may take 30 minutes; that may take an hour. So, when we add all of these times up, there is an hour a day to get ready, an hour a day driving to and from work, and eight and a half hours per day that are tied up in doing what it is that you do on any given day.

The total of all of this is that at least 55 hours of our week are spent in activities related to what we do for a living, and if we are awake for 16 hours per day (112 per week), the quick math shows us that work-related activities account for at least half of a person's waking hours in any given week. If you work a 50-hour week, as many professionals do, then 58% of your life is spent around the work that provides your living. Therefore, the key to a "life-changing transition," as we like to describe it, is making sure that each of us has as many things involved in our work that make us happy and get us excited as possible so we can spend the vast majority of our time doing things that give us energy instead of things that take it away.

The Balancing Act

Most people like to believe that they make logical decisions, but human beings are not logical creatures; we are emotional creatures. Emotion is the part of us that makes us human and not just biological machines. As a consequence, most people make decisions that help them feel better in the moment and then justify those decisions with pieces of logic. This is called confirmation bias, and it's dangerous, as too much of it leads to delusions.

Good decisions align with clearly defined and understood value systems, and value systems provide the basis for directing growth. For example, management may make a decision that seems to align well with value systems on their financial statements, but if the company fails to recognize that the retention and motivation of the people in the business are equally critical, it may find itself making decisions that support one value system while tearing down another.

The same can happen in our lives. Choosing to prioritize relationships over work may fill an emotional need, but without balancing emotional needs with the resources required to live, it becomes self-defeating. Both value systems are important. Developing decision-making skills that maintain balance among competing value systems will help create a happy and successful life.

Every decision that we make opens an opportunity for us in the direction we choose and, by extension, closes off another. Therefore, every decision has an upside and a downside.

For example, let's go back to the location decision. If we make location the primary factor in our transition decisions, it cuts off every opportunity not within commuting distance of that area, regardless of how good it is. We have limited our market for our services to a very limited area. Once that decision has been fixed, another potential downside trade-off may be significantly less compensation than in other areas, or the type of work available may not fit well with our skills and desires. Finally, the worst trade-off can be that it may take more time to find a job in that area, leading to periods of extended unemployment. This is why balancing emotion and logic in making decisions and articulating your value systems stands out as critical.

On occasion, we encounter people whose location choice was driven by a commitment to a family member whose health condition required personal care. This is a noble reason to make the choice, but even though we control the decisions we make, we do not have control over the implications that come from those decisions. Determining whether the implications of our decisions are worth the price is an individual decision. Making a conscious choice to accept the consequences is what makes the decision mature and well-considered.

As a general rule, becoming more restricted in one area of your search requires you to be more flexible in other areas to ensure that trade-offs for your choices do not force us backward in your career progress, assuming career progress is a core value in your system. When forced into having to make the military-to-civilian transition within a very limited period of time, a small percentage of people are able to acquire exactly what they want when it comes to location, compensation, and job requirements. That doesn't mean that all of these cannot be had, but it may take more time and one or more interim steps to reach the goal. Therefore, estimating the time it will take to get to the place you want to be is the next most important element in the transition planning and decision-making process.

As always, the key is balance. If the goal is high compensation and you've decided you want to make as much money as possible, then location and job requirements become the flexible factors in achieving balance. Understandably, maintaining equivalency to military incomes stands out as a primary objective for most people going through the civilian transition, but care needs to be applied in balancing that core value against all the other potential limiting factors.

The Path Already Taken

Consider what you have already done. Joining the armed forces has required you to be geographically flexible, travel extensively, work considerable overtime, maintain a healthy lifestyle, wear a uniform, and work under a contract. In return for this, you have been provided training at no cost to you, a very good monthly income relative to your civilian peers the same age, exceptional benefits in almost every category of life, quality of life resources, and tax advantages.

The military has provided all of this to you because military service demands that you be willing and able to do all that is required of you. Pay rates jump significantly near the end of initial service commitments (E-5 or O-3) because the costs of training to basic proficiency have already been absorbed, and the government understands that replacing you means starting over with a new recruit who requires a bigger investment. Understandably, the goal is to retain you (or at least a certain percentage of people like you), and as of this writing, the government offers a very competitive package to keep you committed, ready, and willing to meet the demands it places on you, including forced relocations, long periods away from home, and above-average working hours.

If you are able to get an offer that meets the government's proposal without having to accept that kind of lifestyle, that is great, but not everyone's skill set will allow them to reach that level of compensation without continuing to do the kinds of things the military requires. Therefore, making a plan that takes these facts into account is smart, even if it proves unnecessary.

By the time we start planning a military-to-civilian transition, most of our qualifications to acquire our next job have already been fixed, and not every qualification developed on and before active duty is equally marketable and valuable. Beyond that, very few have the ability to develop a new, highly marketable qualification in the last six to 12 months of active duty. This is why seeing your civilian transition as only the first step in a career track that will continue for most of your life sets the right expectations. As long as you know where you are starting, a series of decisions that move you closer to your ultimate goal will get you where you want to go, and there are still no limits on how far you can go except as determined by your decisions and the time you have.

When I hear veterans talk about the challenges of the civilian transition, their fear of the unknown and not being able to live as they have tops the list. I wish I could tell you that everyone who leaves active duty will be able to make the transition at the same or greater income level they enjoyed, but I cannot. Military pay has outpaced civilian pay levels at a remarkably fast pace from 2000 through the time of this writing in 2024, and the fact that every military member is paid according to rank and years on the same chart with the same relative promotion timelines for every specialty has created a system that makes maintaining one's lifestyle as a civilian much more challenging.

Though the reality of this may be difficult for your personal situation, the data itself is objective truth, and how one reacts to this truth makes all the difference. The unwise or dumb decision category fails to take into account the reality and truth behind these trade-offs, so avoid boxing yourself into a corner by forcing poorly considered decisions driven by emotions alone.

Helping transitioning veterans understand the civilian work environment so that they can make smart decisions ranks highly among the several purposes I had for writing this book. As your guide, I want to help you create success using commonsense approaches while steering you away from the pitfalls and major mistakes I have seen others make. It is important to understand that I don't control the environment you're entering. I only work to help you navigate it with success.

For example, one of the interesting things that veterans encounter as they're getting out is the reality of actually having to expend time and effort to find somebody to give them work that will pay them at the level they want or need. Top performers in the military typically do not have any issues finding things to do. As a matter of fact, all people in the military typically do not have issues with finding things to do. The military finds lots of things for us to do. That's one of the reasons that we are paid one price. They can work us as much as they want; that's part of the deal.

But when we leave this environment, we quickly realize that finding work that somebody is willing to pay us to do at the level we want requires developing skills we may not already have. Discovering this truth can be a shock to the system, and realizing you don't have any idea how to do this successfully can induce panic, but rest assured that you can learn the skills to be successful. The question is whether you are willing to learn. In other words, are you a professional?

What It Means to Be a "Professional"

Before the 1990s, the term "professional" was reserved mostly for people with college degrees. The idea was that a person who was willing to forgo entering the workforce to earn money after high school and chose to pay a lot of money to pour into work on advanced studies for four more years marked someone serious about his or her vocation. The idea behind all of this was that the professional was willing not only to work for free but to pay for the privilege of working for free. That is effectively what all post-secondary (after high school) education is.

In essence, the people who were selected to go to college, willing to pay for their continued education, and dedicated to completing their degree thought at a higher level than others and, therefore, would be capable of thinking and communicating at levels necessary to be successful in management or technical roles in the sciences or engineering. There is a strong case for this line of thought, but it assumes that university degree programs are the only kind of post-secondary education, and this is what has changed. Self-improvement through a variety of methods was a hallmark of our Founding Fathers, and the key was a commitment to growing one's abilities. Today's culture in America has been embracing this return to a wider view of adult education.

When we pay for the privilege of working for free with the idea that the time, money, and effort we put into our studies will open the door to something better, it seems natural to think, "I paid the price that others are unwilling to pay, so I should be able to live the life that others are unable to live." By extension, the nature and image of a professional is a person who is willing to invest time, effort, and money into developing knowledge and skills related to their vocation. As the world has continued to develop and adult education has become increasingly more common and encouraged, the idea of the professional has moved beyond simply acquiring a bachelor's degree after high school. It has broadened over the last 30-plus years to now be defined by how one manages a career and pursues lifelong learning.

That is why it's important to determine how you see yourself. Everybody is a professional at something. A professional attitude grows from pride and satisfaction in what you do, and everybody has pride in something they do that gives them enjoyment and drives them to become better at it. They prioritize time to do these things regardless of what they are required to do.

Here's an example from my experience. For a number of years, I was a high school football official. At every local chapter of officials within the state, two people are assigned to be the "rules interpreters." These people are responsible for educating the rest of the chapter on how to interpret the rules of the game in every situation, and unsurprisingly, they are usually the most experienced and well-respected members of the chapter. Interpreters have put a lot of time into mastering the rules and mechanics of officiating. At the high school level, they've frequently been selected to work at state championship events, and many of them also pursue their craft at the collegiate level.

One day, we had just finished a chapter meeting in which I'd listened to one of our rules interpreters break down several complex game situations with a level of detail that simply astounded me. He could have stood before any panel of executives in the country with the quality of his presentation and the depth of his knowledge. As we casually sat around the table, enjoying some drinks and food while discussing some of the topics from the meeting, I found myself curious as to what this man did for a living. I said to him, "Wow, this is so incredible. I'm amazed by all that you know about this and how you can present it from so many different angles. You have such a great gift that I have to know: What do you do for a living outside of this?"

Well, the look on his face said it all, and the tone of the conversation soured very, very quickly as he simply told me the name of his employer and turned away. I later learned that this man worked in a bread truck, making deliveries to stores in the local area. He wasn't even the driver of that bread truck; he was actually the assistant to the driver. So, when I shifted the conversation from the work that he was proud to do and where he was highly esteemed to what he did to support himself, I embarrassed him without realizing it.

At that moment, I came to realize the concept of "professional" and professional pride in the modern sense of the word. When it came to football officiating and other local sports, this man was undeniably a professional, highly esteemed in his craft after years of investment and pouring himself into becoming an expert, but on the activity that actually provided his living income, he was not, which is why he was almost insulted to even be asked about it in that setting.

Ultimately, a professional is one whose self-image is tied to whatever work he does. In the case of my friend, the rules interpreter, his professionalism was tied to a part-time gig as a sports official and baseball coach. He could never turn in shoddy work in that space. His pride would not allow it because his self-image was built around his reputation as one of the best.

We see this frequently with veterans on active duty: enlisted soldiers, sailors, airmen, and Marines who can't imagine turning in poor-quality work because doing so would be a poor reflection of themselves that they would never want to see. We see them take great care to protect their reputation, and we see them show up early and stay late to solve problems or take care of other business for their units because they wear their reputations with pride. When we hear these people talk about their work, their stories almost always include off-duty hours when they were bouncing ideas off of peers on how to solve problems or discussing how they would handle certain military situations they might encounter. If this sounds like you, then you, too, are a professional in the work you do in the military.

Again, everybody has something that they pour themselves into that they would do whether somebody paid them. Even being a gamer can be considered a professional endeavor, because many good gamers take pride in the skills that they have developed. Now, career success comes when the focus of our professionalism supports our vocation and the vast majority of time that we spend working in it. That is why identifying with your work is one of the first steps to developing a successful and fulfilling career.

Creating and Navigating a Career You Will Love

So, what is the difference between a job and a career? Ultimately, a job is simply the work we are required to do and/or the responsibilities we maintain. Creating a career involves intentionally linking the succession of jobs we have had to reach the level of responsibility or proficiency we seek.

To use an example from navigation that most of you will understand, imagine that the job we are doing now is like a "fix" on a navigation chart or a map. If this term is not familiar to you, a "fix" is simply your exact location in the world at a particular moment in time. Our current job is where we are right now.

Knowing where we are in the world is important, but it's not enough. We also need to know where we are heading. On a nautical chart, we add our course and speed to the fix to help us understand where we should be in the future. Alone, this would be an arrow coming out of the mark of your fix with its direction and size based on where we are going and how fast we are moving in that direction. In your job, this would be the functional duties and responsibilities we have merged with our current pay rates. These items provide an arrow that indicates where our careers are heading so that we can estimate what should be coming next.

In navigation, we also use a term called "dead reckoning" to describe the process where we use a fix combined with course and speed to estimate our position at two defined points in the future. In open ocean navigation, we dead reckon from our most recent fix to 30 minutes and 60 minutes ahead based upon our course and speed. Applying this to our career management metaphor, your current job and your rate of growth will give you some idea of what your next job should be and when you should move into it, provided you want to optimize the value of the experience you have had. Keeping your career on a steady track (moving in a straight line or as close to it as possible) is obviously the fastest way to get to where you want to go in navigation, and it is the same with your career. Use the value of your past experiences and accomplishments to drive your momentum forward in your current area of expertise.

To further use our navigation metaphor, forces act upon a ship to prevent it from moving exactly to the dead-reckoned position and arriving at the anticipated time. The skill of the helmsman to maintain course and "set and drift" stand out as the most significant factors. Set and drift represent the effects of wind and current on the ship, which alter its true course and speed. The skill of the helmsman to steer also determines how much variance there will be from the intended track.

When we consider these factors in the course of navigating a career, we understand them as both the things that are outside of our control and in our control and how we handle them. Steer yourself well in your current job, and you'll likely arrive at your next job on time. Experience the headwinds of resistance in driving forward, and you may find that reaching your goal on time will require applying more effort. Similarly, a strong wind at your back can get you to your destination faster. So, our skill in navigating the challenges of life and jobs, combined with external factors and our reactions to them, will impact our success and timelines.

When we apply set and drift to our navigation chart, we can see the effects that these outside forces have on us: how we are being blown off course or how outside help is getting us to our destination more quickly. Our skill in steering our course can also shift us a little to the left or right of our track. The key to all of this is taking the time to intentionally look at where we are and how we are progressing and then deciding to either stay on course and hold our speed or make changes.

In terms of our careers, we should be able to look at where we are and where we are headed. By extending that vector to some point in the future, we should be able to estimate the particular role and level of compensation we will have then with some degree of accuracy. So, if growing a career is about increasing your earnings and satisfaction and better using your talents, then staying on the current track makes the most sense.

However, sometimes, things come up where we inadvertently make a turn in our career or find ourselves needing to make a turn. Whatever the reason, we find ourselves taking our career in a different direction. Looking at this from the military-to-civilian transition perspective, we have all gone through a series of progressions that indicate where our military career should be going next. In fact, almost every military member on active duty has a pretty good idea of the kind of assignment that should come next. I've personally never spoken to a military member who didn't know the basics of what the next two jobs in their career ladder should be.

If the goal is to be promoted quickly, then staying on that track is the best way to do it, but the civilian transition represents a point where we all make some type of turn in our careers. Making a career turn is similar to making a navigation turn. With a ship or aircraft, maintaining speed through a turn requires more power than continuing straight on course. The law of inertia always applies—even in our careers.

The Law of Inertia in Making Course Changes

Our internal inertia drives us to keep moving in the direction we are already heading, creating momentum. When life is good, momentum keeps us moving and resistant to change, but if we consciously decide it's time to make a change in our careers and that it's time to leave the service to go in a different direction, we have to expect that there's going to be some resistance as part of that process because our desire for change is working against our momentum.

We're trying to move away from what is expected, and the resistance we feel in our careers is not one of speed, movement, or power applied to our thrust. Instead, it is felt in terms of its impact on time and money—time to make the change and the money that making the change costs until momentum can be built in the new direction. Thus, answering the question, "How much time and money will my intended change cost?" makes great sense in planning the transition.

Practically speaking, changing the course of a career requires more time than other changes and incurs some type of monetary cost —either in the form of a reduction in starting compensation or in an investment in new qualifications. The time involved can include searching for a job that puts you on a new career track and developing a new set of skills prior to making the change.

The importance of this concept to the military-to-civilian transition cannot be overstated. Our time to transition is limited, and we're moving into a world that doesn't do all the same things that we do on active duty. Many of our skills and experiences have relevance, but usually, they require at least some change in how they are applied. After all, not many civilians are flying fighter jets or firing artillery batteries.

Instinctively, we know that we need to align the requirements of our first civilian job as closely as possible to our military job to make the transition easier and maximize our compensation. For most of those making the civilian transition, this is true, but for many others, it is not. This is why there are TAP tracks for full-time school and owning your own business. Pursuing either of these tracks necessitates a commitment to investing considerable time and money in making a major change to your career direction.

The truth is that *everyone wants more*. In our work at Absolutely American, we've twice tested the theory to the extreme that you can never be offered too much, and in both cases, the candidates still accepted the role once they got over their shock. Clearly, these situations are not the norm, but if getting more from your work is part of your plan and value system, I have good news for you. It can be done, and the process is not complicated. Achieving your goal to earn more simply requires a little bit of thought and intention.

Adding intentional thought to making smart decisions and intention to your work will create the kind of career you want, allowing you to retain the things you loved most about military service while eliminating the things you liked least. This will enable you to live not just a good life, but a great one.

Summary:

Remember the first steps for planning your military-to-civilian transition:

1. Be encouraged! The freedom we have served to protect has provided great opportunity for you, and the military-to-civilian transition is only the first step in that journey. A little thought and planning will pay massive dividends in earnings, happiness, and long-term satisfaction. The results can truly be life-changing.

2. Identify your value systems and prioritize which ones need to be satisfied as part of the transition and which ones can wait until your career is better on track—values related to money, family, freedom, location, relationships, work-life balance, and more. If there is a spouse or significant other in your life, you should both do this

independently and work together on setting priorities. Avoid thinking about specific jobs, and make sure that both of your priorities are balanced if the goal is for both of you to be happy in the long term.

3. Evaluate your budget. Look for opportunities to create savings that will enable you to increase your options for your transition rather than limiting them. Buying a house near the end of your time in service can limit your options and create financial pressure more than anything else.

4. Examine the functional duties you performed on active duty. Identify which ones you would like to do more, what you would like to add to those duties that you did not get to do, and determine which ones you would like to eliminate from your civilian career. This will help you isolate the job responsibilities that are the best fit for you.

5. Remember the path you've already taken. You've already benefitted greatly from making sacrifices of location and job requirements. The military-to-civilian transition is the most challenging job change you will make, but opening parameters of your search will lead to more opportunities and a more successful and less stressful transition process.

6. Career inertia works for us and against us. The more change you want to make in a shorter period of time will be felt in the time and effort it takes to make that change and the money that it costs to get there. Money can be felt in periods of unemployment and in costs of funding education or training. Getting a career tracking properly will make career inertia work for you.

7. We are all professionals at something because we all have things we do happily for no money. If you can steer your mind towards being a professional at the thing that also provides your living, you are likely to have a successful and fulfilling career.

Chapter 2
Building the Battle Plan: Define Your Objectives

No battle is ever won without some type of planning. When it comes to executing an On-Target & On-Time ™ military-to-civilian career transition, our objective no longer includes defeating an enemy navy or taking control of a part of another nation's territory. No, the objective of the civilian transition plan is to make sure that our next role is where we want to be, on track to where we want to go, and within time to make sure we can support our habits of eating and living indoors. In everyday speech, we call this a goal, and setting a goal is the first step in building a plan. But where do goals come from?

As discussed in the previous chapter, knowing your value systems is the single most important thing you can do to prepare for a happy and successful life. Defining your value systems helps you understand the direction you want to go with your life. When you know the direction you are heading, you can create a vision for what your future holds, and from your vision comes the goals you set.

Any goal you set for yourself should motivate you. If it doesn't, you won't accomplish it if it requires significant time or effort. Setting goals for your civilian transition is vitally important to creating a successful civilian transition because it provides both a focus for your efforts and a set of criteria to evaluate your success.

The SMART Plan

Long ago, someone developed the little mnemonic device called "SMART" to guide people in putting together motivating goals, and it's been taught as part of a number of military courses, including the Transition Assistance Program, as far back as any living veteran I've met could recall. Plenty of resources exist for using SMART goals, including our Corporate Gray website (www.corporategray.com), but as a brief introduction to this concept, here is what SMART means to the goal-setting process:

- "S" means that the goal you set has to be "**specific.**" If the goal is not specific, achieving it will be unclear, which means it won't happen. A good goal defines victory clearly to any observer.
- "M" means that your goal has to have some way to "**measure**" it. Measuring your progress provides vital feedback on your work in progress and lets you know with certainty when you have arrived.
- "A" means "**attainable.**" This has to do with your beliefs. If you believe that something is attainable, then it is. If you believe that something is not attainable, then it isn't. Strongly attainable goals usually have proof sources from your past that build your confidence.
- "R" means that your goal has to be "**relevant.**" In the civilian transition, relevant means that your goal fits with things that are important to your value system in making this change. The more relevant the goal is, the more it will motivate you.
- "T" means that your goal has to have a "**time**" limit. Time is measurable by itself, but a goal without a time limit is a

"someday" wish that creates no sense of urgency to make something happen in the present. Every one of us has been given a fixed number of days to accomplish everything we will do in this world. Time will eventually run out for all of us, but using time in our goal-setting will help us make the most of our days in this world.

Regardless of the goals you set for your civilian transition, following this model will help you get what you really want. Your goals have to be defined with specific criteria so you can know when you have attained them.

Far too often, we hear in our work that the primary goal for a person's transition is: "I just want to be happy." Happiness, though it may seem like a goal because it appeals to the nature of human beings, is not actually a goal by itself. Happiness is the byproduct of doing the kind of work that brings out our best and progressive attainment of worthwhile goals. Human beings become happy when they feel accepted, esteemed, and safe in whatever they are doing, but the work they do provides the feedback that builds that emotion, frequently because they're successful at it and because it challenges them. Getting to that point requires knowing the work that fits best with who you are and planning how to get it.

In this chapter, we are going to focus on defining your objectives so you can turn them into goals. As part of your transition, you will need to define your objectives and set goals in the following areas:

1. Intentionally defining your value systems.
2. Using your value systems to create a long-term vision for your work life.
3. Establishing the primary objective for your transition that gets you on your path: finding a job, going to school, or starting a business.
4. Determining both the most ideal and minimally acceptable initial results in your primary objective.

Defining Your Value Systems

If you're a person of faith, as I am, you know that work is a bedrock component of faith and is even found in the Ten Commandments. Many people have at least heard the commandment that says, "Remember the Sabbath and keep it holy," and think that is the whole commandment. However, it has a follow-on portion that is often overlooked: "Six days you shall labor and do all of your work." This stands out as unique among the few commandments that require a person to actually do something, rather than prohibit an action. In the context of building a life outside of the military, activities surrounding our work account for more time than anything else we do, and our attitude toward our work makes one of the biggest contributions to our happiness.

When we expect our work to align with our talents, skills, and interests, we pursue it with a positive attitude that makes the idea of doing it enjoyable, whether it's three, four, five, or even six days a week. So, when we experience the positive emotional feedback that we get from doing things that we enjoy, it is easy to see how the right kind of work contributes to happiness. You just can't find happiness by seeking happiness. You find happiness by seeking things that are worthwhile and align with your value system, and happiness grows as a byproduct of that pursuit.

This is why we discussed in the previous chapter the importance of establishing your value systems and how to use them to design the ideal description of the work functions and environments that fit you. When we look at how we impact the world, sometimes we place too much emphasis on comparing ourselves to the big world-changers who have impacted society in ways that we read in history books and the news. But every one of us impacts this world by the countless encounters we have with other people while living our lives every day.

If your work gives you energy and satisfaction, you are likely making positive impacts on this world all the time. The testimony of who we are is measured by the things we do, the things we say, the ideas we cultivate, and our attitudes toward them all. So, when making the plan for your transition, it's vitally important to understand how connecting your work and your value system contributes to building happiness for yourself and spreading it to the world.

Creating Your Long-Term Vision

So, how do you make decisions when you don't yet have answers to the most important questions about your military-to-civilian transition? As I discussed in the Introduction when I described my experiences with mathematical decision theory, it's frequently easier to identify and eliminate the elements of a job and your transition that you absolutely don't want and then realize that what you want to do is at least somewhere on the opposite side.

At the same time, we have to remember that no job will ever be perfect in itself because nothing in this world is perfect. That is why the approach to decision theory I discussed does not have a single "discreet" answer. There is only a variety of options containing the "most correct" answers, which is something that fascinates me about the way math and science help us understand the world around us more deeply. Every job will have elements that are less than ideal for you. This is a consequence of living in a flawed world, but the value of the process comes from constantly pursuing the addition of elements of a job that move you closer to your "ideal" without losing the elements that align with your most important values.

A phrase that I have become accustomed to hearing lately is FOMO. If you are older than 35, FOMO simply means "fear of missing out." A key value among a large and growing part of the younger generation is enjoying life experiences, and opposing that value is the fear of missing out on something good. For this part of the population, life is not nearly as much about acquisition but experience. So, when looking at your civilian transition, what kinds of experiences line up with your value systems, and which ones do not?

For example:

- Do you like to travel? Would you like it as part of a job?
- Do you like working or being off during the day?
- Do you prefer working independently or with a group?
- Do you like city living, or do you prefer small towns?
- Do you like structured or free-thinking work?
- Do you like the idea of living in different places or staying in one place?
- Do you like working inside or outside?
- Is living on a third of what you have been making OK?
- Do you like analytical or creative work?
- Do you like meeting new people all the time or prefer to stay to yourself?
- Do you like being part of the government and its missions?
- Do you like the idea of spending your days learning in a classroom?
- Which is worse: not having a guaranteed income or not having the freedom to do what you want?
- Do you like being around new technologies or learning new things?
- Do you like working with your hands?

These are just a few of the kinds of questions to ask yourself to start narrowing the field to find your path. If you look at my list, you will see that these types of experience questions are designed to help you identify which transition track suits you best—job, school/training, or business.

Establishing Your Primary Objective

One of the greatest gifts that we have in the modern world is the internet. The internet provides us with information at our fingertips, giving us many different options and ideas on what to look at. In most cases, it's a blessing that there's so much information out there that can be found so easily. On the other side of that, it can be a bit over-whelming if you don't know how to filter it.

This is where your values come into determining your plan. If you have decided that you are going on the employment track, then determining the kinds of work functions that give you energy and understanding why they do are important parts of your decision process.

For example, if you're coming out of the Navy, the likelihood of your being stationed in Denver, Colorado, is not very high, but if a connection with the unique features of the Rocky Mountain area is part of your personal value system, that location can be a significant driver for you on choosing to live there. However, that alone doesn't necessarily mean that it has to be Denver itself, even if that is your hometown. No area of the country is unique when it comes to every single factor. There are several options that you can look at that impact part of the location value system. Maybe it is family and being able to raise children around them that is important to you. That certainly affected my choices and explains where I am located right now.

In stacking the factors that are important to you, it's also critical to put them in order of importance. As you start, you're going to be able to define the objective that you are trying to reach. I will give you a personal example. As I already told you, my objective was to make sure that I ended up in a role that had significant income potential beyond just putting time into a job. That's how I found my way into sales and entrepreneurship. But my wife and I are both from the same hometown. In fact, she and my oldest sister were very close friends growing up, and most of our family was located in central Pennsylvania. So, part of my transition involved tying sales and entre-preneurship for my work and income to returning to our home area in central Pennsylvania, where there are far fewer opportunities than can be found in the larger cities around the country.

In fact, when I met with my recruiter rep after I had just submitted my letter of resignation from the Navy and told him about my location objectives, his response was that I had better get used to working weekends because that's what it would take to get me back to where I was. To be quite honest, I was very put off by that because I was willing to do things that I felt other people weren't willing to do to get what I wanted.

So, my personal objective was to get my family back to central Pennsylvania with my working in a role that offered me uncapped income opportunities. That would allow my children to live near and get to know other family members. If getting such a position meant that I had to travel a little bit more, work a different kind of schedule, or accept more risk, my value system said that was fine.

Ideal and Acceptable

Is that how it worked out?

In the end, it did, yes, but it proved to be a two-step process. The first step was that I had to find a role that I could take with me. When it came to living in my hometown area, my feeling at the time was that I would either have to import my job or create it.

Since then, I have done both, but it started with importing. It just so happened that the company that the recruiter rep worked for ultimately hired me to do recruiting in the Norfolk area where I was last stationed. From the work that I did there, I created the success that gave me the ability to transition to work remotely by putting my employer in a position to be very hesitant to turn me down.

Though working from home is very common today, in 2003, it was not. It was very new and not that common at all, but that's what I did to reach my personal objective, which had two components: being around family and having a job that would allow me to make enough money for my needs and my family's needs. This is how you'll plan your objective. You'll start with your first two or three values, and then you're going to connect them to where it is that you want to be one to two steps in the future and what that will look like.

There was one other component that I had to address in my transition—money. The role I accepted was 100% commission-based, and coming from active duty as an O-3 with eight years, I was making a civilian-equivalent income of about $65k per year in 1999. Plus, I had a wife, a child, a house, 2 cars, and two pets. Failure was not an option, and I had to figure out how to make the money work if I wanted my plan to work.

As part of my budgeting process, I knew that taking this kind of role would require driving my expenses as low as possible. I had some debt at the time, but I owned my house and couldn't reduce those expenses. I took a debt consolidation loan to lower my mandatory monthly expenses and was able to refinance a car and some credit cards to reduce my total debt load to just slightly more than my car payment alone had been. In the end, I reduced my mandatory debt payments by nearly two-thirds, which helped me stretch my money further.

I also worked with my command to extend my terminal leave by taking it Monday through Friday for the first two months. Because I was staying in the local area and was able to come in periodically, the command approved it and extended my terminal leave by three weeks. Note: My ability to accomplish this part of my plan connected to planning my transition from a shore duty station where time off was more flexible than a deployable unit, which I had planned and executed years prior to starting my transition process

Between my budgeting efforts, extended terminal leave, and the willingness of the firm to let me start part-time, I was able to fund my life for five months while pursuing my new endeavor with everything I could muster. After doing my due diligence, I believed the plan would work, and it did. In the end, my last full Navy paycheck came to me on May 6, and my first civilian paycheck in my 100% commission job came on May 15.

Serving in today's Navy, I likely would have been able to accomplish all of this by just helping that firm establish a DoD SkillBridge program—another transition option for today's service members that did not exist even 10 years ago. Though what I did is not likely what you are seeking to do, the key takeaway from this is to apply initiative and creative thought to planning and executing your transition to get what you want.

If you're not sure how to get where you want, the exercise from Chapter 1 on defining the functional duties that give you energy and the description of the decision theory on figuring out what you want to do by looking at the opposite of what you don't want to do are worth reviewing.

So, whether you intend to find a job, go to school, or start a business, set your objectives and goals to align with your vision and value systems, and you will be well on your way to a military-to-civilian transition that is on target and on time for you.

Chapter 3
Move from Your Objective to a Plan You Can Execute

D istance to the objective and time available for travel are the first considerations in building any navigation plan. In Chapter 2, we covered setting an objective for your career based on the vision you have for yourself. This vision came from your value systems, and by looking at where those value systems were pointing you, you have picked a place where you want your career to go. Now we are making a plan to get there, and that starts with knowing where you are in relation to where you are going and the direction you are currently heading.

This is where time becomes a factor. Not every objective can be accomplished in a single step, particularly when you have limited time to make that first step. If your current heading is not pointing toward your objective, part of the time you will spend in the military-to-civilian transition will be altering your course to move in that direction. If your heading is pointed in the direction of your objective, you will not have to change course, but you will still have to figure out how close you can get to your objective in the first step and build your transition plan around that. These are called "interim objec-tives," and for the purposes of our discussion, we will continue with our navigation metaphor and call these interim objectives "waypoints."

In navigation, waypoints typically mark a position along a track where something changes. Some waypoints include a stop; others may simply require a change in course or speed. Such is the case with your career. Each waypoint marks a stop along the track to your career objective. Continuing our metaphor, the stop at a waypoint provides an opportunity for you to refuel and improve your career vehicle. Practically speaking, your waypoint will be a job along your career track that gives you the chance to earn while you build qualifications and a record of accomplishment to get you closer to your objective. In the case of education, the stop gives you time to focus on investing in qualifications that can help you change your course and speed to shorten your timeline to reach your objective.

In our search business, we use the rule that any job we present must make a person more valuable to his current employer or other employers on his career track after two years of working in the role. Otherwise, it's not considered a good opportunity.

Time is the fuel of the job search, because time is money in the military pay system. In travel, the amount of fuel we carry limits how far we can move toward our objective before we need to refuel. Time limitations are like fuel and are essential to consider when planning the military-to-civilian transition. When fuel runs low, panic sets in— just ask any aviator about that—but with good planning that takes your track toward a large number of refueling options, fuel limitations can be managed easily and effectively.

Just as with our cars, the key to managing fuel is not pushing the limits of how far you can go but rather making sure you have lots of refueling options available. Such is the case with your job search, which is why we encourage those in transition to align the parameters of their searches in the direction of the greatest number of job options when time is most critically limited. During this period, your military pay is what keeps your job search engine running, and your pay is directly tied to your remaining time in service. That is why time is like fuel. Though you may have some reserve fuel in another tank (such as savings, which is good) or a backup supply (such as a working spouse or investments that provide cash flow), the military pay that comes every first and 15th is almost always the largest source of income for a service member in transition.

Although a straight line is always the shortest distance between two points, the waypoints on your career track in the real world are not likely to be in a perfectly straight line. If we look at time in the job search like fuel for an aircraft, the airports where you can stop to refuel and overhaul are where they are and not necessarily right along your track. Many times, aircraft have to go considerably off track to find airports where they can refuel and/or overhaul.

Such is the case with jobs. The jobs available along your track are where they are. They may not be exactly on your track, but they are able to be found in range of your fuel (time). Though this may not seem as daunting in land and sea navigation, aviators know better than anybody else the importance of constantly knowing your fuel status in relation to your objective. Regardless, you will need a place to stop within the limited range of your fuel, even if the stop takes you a little off track. That means that each of your waypoints should give you the chance to make course and speed adjustments.

The purpose of this book is not to help you manage your entire career. It is for managing the military-to-civilian transition portion of it, but I hope that visualizing your career this way helps you better understand it in its broader context. One of the most important things that you have to understand is that every plan has a timeline, and we've already discussed how time is of the essence in the military transition process. There's a definite day when your pay is going to stop, so the plan for your transition has to take into account your timeline and the narrow window you have to execute your plan.

Timing Matters

Think of this transition as a mission. There is a time when it is too early for you to execute the mission. You can plan and rehearse, but until the moment is right to execute, you have to hold back. Such is the case with the job search. You can certainly do all the preparation and planning as we have discussed. You can even start networking well in advance of your availability date, but when it comes to executing your active job search, you have to focus all of your efforts on the sweet spot in time to be most effective because very few employers will be able to hire you while you are still several months away from the date you can start.

Similarly, you can find yourself too close to the end of your timeline, trying to find a job when you're too close to your end-of-service date to secure a role you really want before you stop getting paid by the service. A mission executed too late runs into far more resistance. Though it can still succeed, it is frequently harder and costs more.

If your unit is deployed when it's time for you to get out, you may find yourself in this position in spite of your best efforts to prepare. If that is your situation, you have to deal with what's in front of you. No matter what you do, it takes time to generate leads for your search, go through the interview process, get an offer you can accept, and complete the pre-employment checks before you can get started in your new role and get paid again. If you think you may be deployed when you are scheduled to transition, I highly recommend saving all the leave and money you can to give yourself as much time as possible to execute your search without a negative financial impact. Nowadays, the good news is that even deployed units have access to the internet and resources to enable you to make your transition plan while you're overseas.

Most people coming off of active-duty military service are not hired more than three months from their date of availability. Quite frankly, the typical date is right around 45 days, with a range that generally falls between 15 and 60 days. If you are an officer or come from an extremely high-demand enlisted specialty, you could be hired up to 90 to 120 days out, but for most service members, 30 to 45 days from availability remains the sweet spot. These timelines have nothing to do with the service members and everything to do with the timelines employers use to fill the kinds of roles that align with the Day One capabilities of most transitioning service members.

Another driver of the timelines with employers is relocation. Most service members will need to relocate to accept their first position after active duty. Even in the civilian world, companies will typically allow 30 to 60 days from acceptance of an offer for people to relocate as part of starting a new position. So, including time to relocate has to be factored into most transition plans, which means building your plan around getting your ideal offers between 15 and 60 days from your intended start date for best results.

This is actually very doable, and over the course of my career, I have helped thousands of veterans find a job that fits their value systems within this timeframe. Executing a successful military-to-civilian transition requires having faith that there is enough opportunity in the world to meet your goals and that there will be for you as long as you build your transition plan with enough flexibility to allow for a relatively short time to execute a search. That means steering toward the largest number of opportunities for you.

You can rest assured that the need for help abounds all over this world. There is no lack of appetite in companies across America and in the government for people who need good help. There's never enough, and there never has been. If you have been following the thought thread of this book up to this point, you may be starting to understand why.

"Help Me Help You"

This is one of the great lines from *Jerry Maguire*, and after accounting for time, it's the mindset shift that is necessary to create a plan that helps you reach your objective. As I said previously, everybody is looking for help, and at the core of all we're doing, the central theme to remember is that the sign says, "Help Wanted." So, before we go into some of the specifics of the plan for you to find the role you want, it makes sense for us to talk about where the jobs come from.

As you're going through this process, you're probably experiencing a lot of discomfort. Discomfort that comes from the pressure of having to find a new role and make a move that will satisfy you and your family's needs in terms of what you are going to be doing, where you're going to be doing it, how much you're going to be paid for it, and what your lifestyle is going to be like as a part of that process. However, the employer that you're initially talking to isn't really concerned about any of that, at least at the beginning. They have their own list of things that are important to them. They have a business problem that needs a people solution.

If a job is posted, that employer has already figured out that the goals they need to accomplish cannot be achieved with their existing staff, which is causing them significant pain. But the other source of pain grows from the process of having to hire and onboard a new employee. To satisfy the requirements of their responsibilities, they have to go through the process of identifying candidates, interviewing, making offers, and onboarding. These processes introduce a whole new set of pain to the manager.

As you can imagine, the employer will only actually go through the pain of the hiring and onboarding process if the pain of keeping the position open exceeds the pain associated with the process of filling it. You may be surprised to learn that civilian managers don't get any extra time in their lives to be able to go out and find people to fill the positions that they have for their teams. Any work they do in recruiting for their staff is added to the work they already have to do with not enough staff. Larger companies will hire human resources and talent acquisition professionals to help managers find appropriate candidates to hire, but the manager alone bears the burden for positions that remain open.

That's an important concept for the savvy job-seeker to understand. When you understand your real audience and what your audience wants, it becomes much simpler to understand what is expected of you. To use a metaphor, the goal for every job-seeker is ultimately to be the anesthesiologist, or the pain pill, for the manager who has a hiring problem that is causing them pain, which is the essence of the plan you are building. **Your plan when reaching out to employers is *not* to talk about your needs but to introduce employers to your abilities and interest in taking care of *their* needs.** Satisfying your needs will come as a direct result of your efforts to take care of them.

In the last half of the 20th century, Zig Ziglar became world-renowned for his sales and mindset training. He sold more books, tapes, and in-person appearances than any other sales trainers in his era. From listening to one of his recordings, I remember hearing him say that there is only one radio station in the world that everyone listens to: WII-FM, which he defined as "**What's In It For Me?**" He went on to describe how all effectiveness in influencing people's thoughts and actions comes down to positioning everything from the perspective of how it will benefit the other person.

Obviously, you are interested in making a successful military-to-civilian transition as you've defined it according to your values, but managers talking to you or reading your application only care about the problem that they need a person to solve. To convince a manager to hire you, you will first need to get their attention to book an interview with you, and during the interview, they will need to see you as the solution to her problem.

At this point, you may be wondering, *How do I do that?*

Imagine for a moment that you are an employer who needs a person to solve a particular problem. You're taking the time to comb through the hundred resumes and applications in your queue because you're under increasing pressure to hire someone who will fill the need, but you don't want to hire the wrong person who could cause you problems or quit right away. After all, you don't want to go through all of this again if you can help it, and nobody wants to hire more problems. With this in mind, what would get your attention if you were going through a stack of resumes on your desk?

I'll bet you would start by finding all the people who could do the job by identifying those with at least the basic qualifications you need. How would you do that? Well, by filtering your applications based on the most important criteria first. For example, if the role to be filled required skills in "maintenance," you would likely filter the applications using the keyword "maintenance."

Let's say that our filter takes our hundred resumes down to 62. That's good, but it's still too many to review in detail. The next most important functional duty may be "troubleshooting," so you filter further using that keyword. Now, you have 39 applications. Then you say that you need someone with "electrical" experience, and you filter further, leaving you with only 19 applications. That list is small enough to review by hand.

Once you start into this group of 19 applicants with maintenance, troubleshooting, and electrical backgrounds, what would you look at next to determine the seven to 10 candidates worth meeting? I suspect you would start looking at the details of their experience for relevance to your work environment.

Let's say this takes you down to 10 applicants you intend to interview. The problem is that your boss just called you and needs you to address an urgent issue, and you realize that you will only have time to reach out to three candidates. Which of the 10 do you call? What is going to get your attention?

When I ask people this question, I always get the answer that everyone is trained to give: "I would assume that the manager would look for the most qualified candidates and call them first." Though that may be the way this manager starts the process, it won't be how they finish. The candidates that get attention are going to be the ones who seem to have experience solving the kinds of problems that the manager has and seem to really enjoy doing that.

"How will the manager know that?" you ask. It will be in the candidate's file because the candidate who gets the manager excited will be the one who read this book and made it easy to find that information in his file. This candidate will not only talk about all the functional duties related to the job, but in the same package, about the work they want to do and why they like to do it. The candidate will explain the problems she believes can be solved and that she wants to solve.

How did the candidate know what message to send? First, the candidate took the time to know herself and identify the functional duties that provide energy. The candidate then searched for jobs using those key phrases. That created a tightly targeted and manageable list. Then, this candidate took the time to read the job description to understand the manager's issue and made sure that in their outreach, the manager's issues were addressed using the candidate's abilities.

That is one of the first things you need to think about as you move from your objectives and strengths toward putting together a plan to engage the market. Using the functional duties you like to perform, you will search for positions that describe the kinds of pain points that you like to relieve. This is how you start making the connection between what you want and what the employer wants.

I'll give you an example from my personal experience. About 20 years ago, we were working with a company that was doing a large number of interviews with us, and the role they were trying to fill was very significant to the particular manager who was involved. I don't remember the specifics of the position or even the company at this point; what I do remember is the story of how all of this happened.

During the interview, our candidate was sitting in front of the employer. Throughout the course of the 30-minute interview, the employer's phone kept ringing, and he had to continue to pardon his interruption from the interview. After taking three or four phone calls while the candidate patiently stood by, the manager looked at him and said. "You know what? I hate to do this. I really hate asking this question because I know you've been sitting here patiently, waiting to talk with me, but all of this happened. Can we just cut to the chase? Tell me, why should I hire you?"

The answer our candidate gave has stuck with me for the last 20 years. He simply said with calm confidence, "If you hire me for this role, I can make that phone stop ringing."

Needless to say, this veteran candidate was hired on the spot, and we've been telling that story ever since because it gets to the heart of what everybody wants. Though this story came from an interview and not a job search plan, I use it here to illustrate the power of focusing the work that you want to do right at the center of somebody else's problem. Even if you don't hit dead-on with what that person needs, you will stand out for trying to be of service, and that will frequently get you a conversation as long as you're somewhat close to the mark.

When I meet new people, the person that I am going to like has an interest in solving problems for me. I am an extremely busy person, and meeting a helpful person is always going to get my attention because I can always use more people like that. So, as you're putting together your plan to engage employers, the first thing you want to be thinking about is how you can use what you like to do to solve problems for somebody else. Then you make your offer.

Planning Your Search for Fun and Profit

Up to this point, everything we have discussed has been focused on defining the things important to you—your value systems, the duties you like to perform, where you want to live, etc. None of this helps you with getting paid until you can do it for somebody who has that particular need. So, now that you understand yourself, let's shift gears and talk about how to start putting together the plan to execute your job search.

Where should you be right now? Well, if we look at the ideal plan, most of the work that will be done as part of your job search is literally happening while you're reading this. You are currently employed

—if you are still on active duty—and that means you have a series of resources that have been provided to you. These resources can include the kind of training you've had, the education you've received, the organization that you're with, tools, publications, people resources, and equipment resources—a huge host of things that you could be responsible for.

Someone is evaluating your performance on how you're taking the resources and responsibilities entrusted to you to meet the expectations of your role every single day. So, the first key to getting your next job starts with performing well in your first job, the job you're in right now. If there's one thing that never goes out of style and that's completely within your control, it's the ability to give your very best and perform at a high level.

If you're already doing that right now and already being recognized for that, congratulations! If you haven't been recognized for that yet, there's no time like the present to decide to elevate your level of performance and focus on delivering more value in what you're doing. One of my business mentors told me long ago, "There is always a place for the person who is willing to do whatever it takes to get the job done and a little bit more." Everybody wants that person on their team.

Assuming that you are reading this book and are about one year away from your earliest date to start work, this would be the ideal time to have the items in Chapters 1 and 2 locked down. You have your value systems identified and prioritized. You have clarity on the functional duties you have enjoyed and would really like in your next job. You have an idea of where you want to live, ideally with some acceptable alternatives. You've already started preparing your budget for the transition, and you're working closely with others affected by the decisions you are going to be making.

You've put together a set of objectives for your transition that make sense because you developed them from your career vision, which came as an extension of your values and the work you like doing.

Taking Your Fix

Now that things are starting to get serious, it's time to take a fix on where you are in relation to your career objectives. This is the starting point. We don't have the benefit of being able to change where we are now, so your plan has to connect today's reality with tomorrow's possibility.

This is where you will begin the inventory of defining where you are right now. Document the training and the education you have. Everything is valuable. Even your initial indoctrination training provides value to you in terms of shaping your character and your abilities to take on a new responsibility.

Document the experiences you've had and all the functional duties you've performed in your various roles in your time in service. If you have more than six years on active duty, put the greatest emphasis on your most recent six years and connect that experience to your early years. If your role involves things such as infantry responsibilities for a small team, your responsibilities relate to caring for the welfare of your people, making sure the organization meets its operational objectives, bringing people through training timelines, and keeping things moving on time.

When it comes to our jobs, we all have a series of functional duties, and we should know what they are and begin to document them. Document any responsibilities you have had for people, property, missions, and programs. Responsibilities imply a trust to meet an organizational goal, and they frequently include supporting resources. Articulate what you accomplished, how you used your resources, and why the things you accomplished are important, if it's not obvious. Performance always sells. If you received an award for performing your duties, it's much more important for you to articulate what you did to get that medal than simply mentioning that you received it.

These are the building blocks you will put together to create a resume that sells. At its essence, your resume is nothing more than an articulation of your education and training, a description of your responsibilities that required you to perform a number of functional duties. From your actions, you have created successes for your organization that have contributed to its purpose. This is the information you need to fix your current position on the chart of your career.

Reading Current Course and Speed

Now we have to look at our course and speed. We're trying to figure out where it is that our current career path is taking us and how fast it is moving us in that direction. Needless to say, if you are still on active duty and reading this book, you are in a particular profession, and the most obvious thing that you would think is that the skills, experiences, and accomplishments that you have are most ideally focused on the military.

That's absolutely true, because that's the environment where you have been working. The military is a highly unique environment compared with other occupations. However, lots of things are done as part of military life that are found everywhere else. For example, the military employs large numbers of people from all different types of backgrounds, from all different parts of the country, and even from different parts of the world. There are people who work in businesses and for the government who come from all different backgrounds and work in all different parts of the country and all different parts of the world. So, that's a similarity.

Here's another: Equipment and technology are found in the military and all other parts of government and business. If you're working on an M-1 tank, it is not much different from working on a conveyor system in a manufacturing plant. And if you work on a radar, the electrons don't behave any differently in a CT scanner. Or perhaps you lead small teams in an infantry unit, where your job is to "take that hill." The motivational techniques that you use every day and the way you lead your people are very similar to the way that people want to be led outside of the military.

Once you understand where this is pointing you and you've looked through how your functional duties, skills, and experience line up, the question to ask yourself is: *Which of those skills, qualifications, experiences, and successes will take me in the direction I want to go?* These are the things you will want to emphasize in building a resume, writing a job search letter/email, writing a cover letter/email, or submitting an application. Check out the resources on Corporate-Gray.com for the most up-to-date information on specifics of how to build your resume.

When it comes to your resume, the most important takeaway from this book is that you understand that the information you add to it has to connect with the direction you want to go. You must understand that your resume is not a biography. It is a sales document masquerading as a biography. The resume shows its reader where you have been, but it has to connect with where you are going. A general resume may get you a job, but it is highly likely that it won't be something that moves you forward in a career. A well-considered resume that emphasizes the work functions you have done and want to continue doing, combined with some type of insight on where you want to go, will eliminate some jobs you could do—but it will also attract more of the job options you really want.

Once you have established the functional duties you like and want, you can use these phrases to search for jobs that fit the kinds of work you want to do. The more options you can find, the more likely it is that you will be successful. There are plenty of commercial job boards, but you may find it easier to use military-to-civilian transition resources such as Corporate Gray to find jobs and employers whose roles are written to be more friendly to veteran job-seekers.

In summary, your key takeaways from this chapter of moving from your objective to a plan that you can execute are as follows:

1. **Know your timeline.** Time is the fuel that drives your job search engine. Accepting your offer between 30 and 45 days from your availability date is the ideal time for most

service members. Build your plan around that timeline if your operational commitments allow it.

2. **Take a fix on where you are** in relation to your career objective. Start planning waypoints to get you from your current role to the role you want to reach.

3. **Build content for your resume** with emphasis on the functional duties you have most enjoyed that you would like to continue in your next role. Describe your education and training in ways that support your interests.

4. **Conduct job research** using different phrases that describe the functional duties you want, and focus your search toward the largest number of opportunities along your career path that can be reached within your limited time. Your research will help you discover where most of the opportunities you want can be found.

5. Use your research to **start making connections between what you want to do in your new role and the help that employers need.** Plan your outreach and applications around introducing the functional duties you most want to perform and how your work in those areas can solve problems for the employer. Remember, the employer first thinks about solving their own problems and relieving their own pain. Bend that to your advantage by addressing it upfront.

6. **Seek military-to-civilian job search resources and job boards.** CorporateGray.com is a great place to start because it contains all kinds of resources for your transition and future job searches, as well as examples of many of the things discussed in this book.

Chapter 4
Finding Your High-Value Targets

I f you're pursuing the job search track as part of the military-to-civilian transition, having to find work seems a strange process for the active duty military member, as I mentioned previously. But if we look at this in terms of a military campaign, no campaign is successful without making sure that you engage your high-value targets. These are the targets that give you the most leverage on what you are trying to accomplish on the path to your objective. Finding high-value targets requires military intelligence, and the same is true for our job search process.

Find a Guide

All good military intelligence involves the use of some type of guide or in-country consultant, and successful military leaders who work in these foreign territories understand the importance of engaging experts who know that particular area. It's more than just knowing the points on the map; it's knowing people, places, cultures, and interactions. Of course, it's also knowing the tripwires and the danger zones and how to avoid the catastrophic failures that they bring. Most importantly, good guides know where the opportunities are, because they're up to date on what's happening in the area.

The value of a good guide in a foreign military operation cannot be overstated, and the same is true with your military-to-civilian transition. After all, this is new territory. As a person who's naturally inclined toward doing things myself, I certainly understand the mindset of the rugged, self-sufficient military member who believes that he or she can figure out a solution to any problem—because it's probably true.

Given enough time, smart and determined people will adapt and overcome obstacles until they reach success. But when it comes to your career, just like it would come to going into court by yourself, I don't recommend it. The cost of failure is too high, and with a short timeline, the odds are not going to be in your favor if you go it alone. Yes, you may be able to find a job on your own, but the likelihood of finding the job that you both fully enjoy and that moves you furthest in the direction you want to go is extremely low.

After all, you're going to make one military-to-civilian transition in your life, but my team has been doing this every day since 1994, and I've been personally helping create successful military-to-civilian transitions since 1999. Over that time, we have successfully helped tens of thousands of military members to navigate successful military-to-civilian transitions, and as you can imagine, we've encountered every transition situation out there dozens of times. Though some of the specifics of the situations may change, the underlying issues never do.

So, what's the point? Ultimately, working with experts who have gone down the path many times before can save you a lot of pain and frustration. Beyond Corporate Gray and Absolutely American, a vast number of transition assistance resources are available to military members, and all of them can be helpful. No, they're not all equal in their quality and capacity, but all offer the expertise that transitioning service members need. However, you have to be willing to seek that help out. So, if you're reading or listening to this book, you can congratulate yourself because you've taken an important step toward finding the answers you need.

If you're looking for a place to start, your base Transition Assistance Program is a great choice, and there are a number of other resources as well. Corporate Gray is a terrific example of a military transition company that serves more than 20,000 military job-seekers per year through in-person and virtual military-friendly job fairs, a job board exclusively for employers seeking to hire veterans, and resource pages with every resource a veteran would need to execute a successful transition. Absolutely American (www.AbsolutelyAmerican.com) represents a great example of a high-quality military recruiting firm. You can also look at veteran support organizations such as the Military Officers Association of America (www.MOAA.org) or the Non-Commissioned Officers Association of America (www.ncoausa.org).

Trade associations for the industry you want to enter can be another great resource, and the vast majority are very friendly to military veterans.

Building a Network

The second tactic for finding your high-value targets is building and expanding your network. When we talk with employers about the sources of their best employees, we have always found that companies are very interested in hiring veterans. They frequently credit veterans as being second only to referrals from existing employees as their best source for successful hires. So, if you're trying to find the targets that are going to work well for you, networking with other people—especially other veterans—is a great way to find ones that fit your skills and interests.

Networking provides a highly effective tool for the job search. In its simplest form, networking uses something common between two people previously unfamiliar with each other as the basis for establishing a new relationship. Generally speaking, the value and longevity of the new relationships are usually a function of the number and relevance of the common experiences or relationships between the two people. The more common points of reference and the more relevant they are, the more likely the relationship will be useful and long-lasting. Therefore, only a few of the relationships you develop through networking will be fruitful for your job search and last for a long time, but the ones that do make the effort more than worthwhile, which is why no career can be developed effectively without networking of some type.

Networking in the current times cannot be discussed without talking about the impact of social media platforms, LinkedIn in particular. At the time of this writing, LinkedIn is far and away the single best job-related networking tool ever created. It allows you to maintain a constant presence in the working world among people you already know or are close to you. The key to succeeding with LinkedIn is connecting with as many people as you can who share your common background and job interests.

Entire books have been written on using LinkedIn successfully and getting jobs and growing careers. If you're reading this and not already using LinkedIn, you're already missing out on opportunities for growth and happiness—and if that doesn't trigger your FOMO responses, nothing will. LinkedIn is unlike other social media platforms because its focus has been on business from its very invention, and you need it, period.

When it comes to building your LinkedIn profile, we recommend that you build it from your resume with an emphasis on the functional duties you would like to perform. Remember, while you're doing that, your branch of service is a single employer, regardless of how long it is that you have been on active duty. You can certainly break down your different jobs on your LinkedIn profile, but be sure to emphasize the work that you've done in the last five years, particularly the functions that you'd like to continue to do more of from the list that you built earlier.

Adding to that, LinkedIn gives you a number of ways that you can tell the world how you can help through the functional duties you want to perform. You can even tell the world about the next step you want in your career, and it operates 24/7/365 for you.

Many veterans in transition never build a LinkedIn profile, and of those who do, many never update it after leaving the service. That is a major mistake. Without knowing what you are doing after the military, nobody in the world has any idea of what the next right step could be for you. If you pay attention at all to your account, you'll continue to get outreaches that only relate to the civilian transition long after you've gotten out, and that's a waste of time for everyone. If you're serious about your career, keep your LinkedIn profile up to date. In the near future, we anticipate that well-prepared LinkedIn profiles will be widely acceptable alternatives to traditional resumes.

Interested in the science of networking? In 1994, three students at Albright College in Reading, Pennsylvania, developed a game called Six Degrees of Kevin Bacon. By the early 2000s, it had become a grassroots phenomenon and was all over the internet. In today's terms, it went "viral," and everyone was talking about their six-degree connections with all kinds of famous people. The game was inspired by the theory of "six degrees of separation," which mathematically demonstrates that every person in the world is six or fewer relationships away from every other person in the world. Kevin Bacon became the namesake of the game when he made a comment during a 1994 interview with *Premier* magazine that he had either worked with everyone in Hollywood or worked with someone who had worked with anyone he hadn't.

If you get good at building your network and talking to people, you won't have to go through all six degrees to get to Kevin Bacon or anyone else. Truth be told, the reason why LinkedIn uses the three-connection step limit is based on the fact that you really should not have to go more than three steps from yourself to find what you're seeking in a personal networking situation. That's smart research.

Then, there is personal networking. This is doing it the old-fashioned way, which can also be enhanced using LinkedIn. Of course, the best time to build a personal network is always before you need it. But if you haven't started to build a network of relationships that can help you, today is the day to start, and here are some ways to prime the pump.

Start with veterans that you know who have already left the service. Every service member has known other people from their units who have left active duty. If you haven't stayed in contact with those people, now is the time to reach out and make contact with them again. When it comes to building a network before you need it, it's important to remember that nobody wants to feel cheapened by a relationship that only matters to the other person when they want something.

So, when it comes to reaching out to the people you've worked with in the past, be interested in what they are doing and in their career transition experiences. One of the defining characteristics of military members and veterans is an almost innate desire to share their experiences, teach other people, and help them along the path that they have been on themselves.

Veterans carry this attribute into their civilian careers. It's one of the reasons they make great employees. Military service builds within us a culture of team and the value that "a rising tide raises all ships." So, when you go into this, being interested in the other person's experiences with their current employer and the whole transition process will get them interested in telling you about them, giving you the insights you need from somebody who has walked the path.

I very much recommend holding off on any asks at this point. Asking for something from someone on a reconnection just rubs people the wrong way. And as I mentioned previously, it cheapens the relationship. It sends the message, "I only care about what you're doing because I want something." Look for ways to build the relationship and add value outside of the job search. After all, that's what friends do.

After those you know personally, you can move on to those who are second-level connections from you (people connected to your friends but who are not connected to you), then move to the third-level connections with a continued focus on staying in the areas that fit your common interests, strengths, and experiences.

Just understand that networking outreaches to former colleagues and connections with more degrees of separation have to start with a genuine interest in the other people. Save your "asks" for a later date. We're going to explore this concept in-depth when we get into the interview process, because the relational part of the interview is very similar to the relational part of networking. Just relax and understand that there will be time, and the truth is that your interest in the other person and their transition experience will already bring your needs to mind without you saying a word. The people you contact may even ask you what you want. If they do, let them take the lead on the subject.

Also, asking for any "lessons learned" is a fundamental part of the military's continuous improvement process, and we use lessons learned in the military to develop improvements to everything we do. Veterans are frequently willing to share what they have learned.

Using Search Engines

Everything in this world nowadays seems to be done on the internet, and research is most easily done there. Since the advent of search engines, the world's access to information has changed incredibly. Certainly, chief among the search engines out there is Google. So, if you want to conduct research to try to find the kinds of jobs that include the function duties you want to do or add to your job, one of the easiest ways to find them is with a simple Google search where you enter the phrases of the functional duties you performed on active duty.

For example, if you like your work supervising three to four people in a warehouse environment and want to continue doing that, simply put "job supervising three to four people in a warehouse environment" into Google and hit "enter," and you will see all kinds of options for warehouse supervisor jobs. You can do the same thing in the search bar of any job site. Internet search engines just have the advantage of being able to touch all kinds of job boards and not just one.

If you're looking for a specific area, just add the location: "Job supervising a team of three to four warehouse personnel in Pittsburgh, PA."

Why is this so important and so effective?

A high-value target for you is based on your desire to perform specific work functions in an environment where you will enjoy doing it. So, if you would like to narrow the shot group of the number of employers you pursue, taking a phrase that you've created yourself and searching for that in a job description is going to lead you to exactly what you're seeking. As you read those descriptions, you will find others that articulate what you want in a slightly different way and give you new phrases to use for your search.

Narrowing the field down from there becomes much, much simpler. As you conduct your job research using each of the functional duties you want and have either performed or want to perform, you're going to find many job postings of interest. Here's the magic.

Most companies' applicant tracking systems use keyword and key phrase matching to prioritize job applicants. So, when your application, cover letter, and resume list the specific functions you want in a job and those align with its highest prioritized functions, your automated matching score will rank you higher than other applicants, giving you a better chance of being contacted for an interview and getting the job.

So, let's turn back to LinkedIn for a moment and discuss job searches. As I mentioned previously, LinkedIn is the single most powerful job search tool that has ever existed. The reason for this is that LinkedIn offers the ability not only to connect with jobs but also with the people who are associated with those jobs.

There are two ways that you can do your research on LinkedIn.

Finding Your High-Value Targets

The first is that LinkedIn has its own job posting pages, and you can use the LinkedIn search criteria to find jobs that match what you're looking for in the same way that you would on Google. Then, you can make an application directly through LinkedIn. This is fairly straightforward, but it is not as effective as the other, less straightforward but far more powerful approach.

Once you've found an employer with the job you want, LinkedIn will also give you the ability to look at profiles of people who work for that company. Those people have their LinkedIn profiles set up with an indication of the town or area where they live. For example, let's say you would like to work for General Electric because you saw a position on LinkedIn in the Cincinnati, Ohio, area that seemed to be a good fit based on the job duties and location. When you go to General Electric's company page, you can use the "people" tab to search for other veterans or specific types of veterans (perhaps directly from your branch of service or even from your specialty). In this case, once you go to General Electric's company page and people tab, you can filter all the people results down to everybody who lives in the Cincinnati area or Ohio and Kentucky.

You can narrow it down further by looking for keywords in the profiles of the individuals in that area. If you put "infantry" in the keyword search block—a keyword unique to military service—it will find all employees in that area who have "infantry" in their background. Finding this common background helps you make a connection with that person, and with some personal networking and relationship building, they will likely want to help you "get in" with the company.

Why? First, it is the nature of veterans to want to help other veterans, but that only goes so far. The next reason is more direct. Many companies offer bonuses to employees who refer good employees to the company. Even if there is no financial reward, employees who are successful in bringing new people to the company are frequently recognized by management for their contributions, which helps

63

improve marks in performance evaluations and increases the opportunity of earning a promotion.

Here is one of the most dramatic personal examples I know of someone having success with this tactic. We will call this person "Dave." Dave never served in the military, but a few years ago, his parents, knowing the kind of work that I do, came to me and asked me to help him find a job if I could. Wanting to help my friends through this "failure to launch" scenario, I agreed to help. I discovered that Dave was intelligent, but he had taken the previous six years to finish his hard engineering degree. Even with the extended time, his GPA was low, but he'd persisted and earned a degree that very few complete. At the time I entered this story, Dave had graduated six months earlier, had no job prospects at all, and had not yet landed a single interview.

I've had several opportunities over my career to help people who were struggling with their job search, and quite frankly, most of the time, the problems were fairly easy to solve. A few misunderstandings about the market, a few tweaks here and there on a resume, a little interview prep, and the person was good to go. That was not the case with Dave. This was the single most challenging case I had ever encountered. The young man was bright and good in appearance, but he did not talk a lot and was very passive. When I first looked at his resume, I saw that he'd given more space to highlighting a part-time job he had as a host at a Cracker Barrel than he did about his six years in the engineering school at a large university well-known for its engineering department. Beyond that, Dave was a bit socially awkward, but he was determined to get a job in his field of study.

After we had gone through the process of discovering his strengths and articulating what Dave really wanted to do, we got his resume improved and focused on his objective. The last step of the process was using his LinkedIn profile to network. I suggested that he look at the different kinds of companies that fit the industry and locations of interest to him. It took a bit to get him to do this, but ultimately, he did. Then I suggested that he reach out to alumni of his school, especially those in the same major, who worked for those companies in his locations of interest.

After six months with no prospects, Dave was able to arrange two interviews for roles he wanted within two weeks of engaging fellow alumni through LinkedIn. With a couple of interview preparation sessions to help him overcome some of his introverted tendencies and bring his passion for engineering work forward, Dave successfully navigated the interview process and was hired into his dream job by a mid-tier defense contractor in the exact location he wanted to be. Though I suspected that Dave would likely work for that company for a very long time, I recently learned that he had just been hired by the federal government directly with a significant pay raise in the same office he had been supporting. He didn't even have to move his desk.

Regardless of your background, you can be sure this approach can work for you, too.

LinkedIn's power to connect you with people who are similar to you in almost any place in the world, particularly in the business community, is unmatched. Using it can be a strategy to help anyone, especially military members, because of the millions and millions of service members who work in industry every day. When they find each other in the workspace, they always come together and always have or almost always have an immediate rapport with each other.

Summary

1. Everyone can benefit from a guide. Seek out resources that can help you find success and avoid pitfalls. Military transition companies like Corporate Gray can help veterans of any rank or specialty. Veteran Support Organizations like MOAA and the NCOA have resources that can be helpful to you, and trade organizations for the industry you want are likely to be friendly to veterans.

2. LinkedIn is your most powerful technology tool. Its reach can help you connect with other military veterans with similar backgrounds and career interests who will likely be willing to connect with you.
3. Reconnect with veterans who have already transitioned. Be interested in them first and hold your asks for later.
4. Use search engines and LinkedIn to search for jobs using phrases from the duties in your resume you want to do in the role you seek.

Chapter 5
Applying, Interviewing, and Evaluating

I n the work that I do in our Absolutely American business unit, one of the things that I'm frequently asked for by the candidates I work with in their transition is a copy of the company's job description for the position we're working to fill. You may be surprised to learn that I'm always a bit hesitant to give the employer-provided job description to the job-seeker. The reason for that is simple: reading a company's job description frequently convinces the military job-seeker that she's not qualified for the job. Having confidence in your qualifications for a position is one of the biggest keys to being successful not only in your application but also in your interviewing and offer discussions.

The good news is that there are a number of things that help establish a person as qualified for a position that you may probably never consider on your own. What are they? Well, we break these down into two primary categories: "literal" and "relatable" qualifications. "Literal" means exactly what it sounds like. The job literally requires you to have a specific qualification from another position or education, and frequently, these are expressed in terms common to the industry. You know this from your own experience with learning military jargon, but literal qualifications are absolutely necessary to meet the minimum competence standards on Day One. The problem is that employers often say that literal qualifications are mandatory in job descriptions when, really, they are not.

If you are applying for a job that is the same or nearly the same as the one you did on active duty, then this would be no problem for you, assuming that the job doesn't require a new license. Military jobs with direct civilian equivalents frequently include military specialties in the medical community. Unfortunately, some military medical professionals struggle with transitioning their roles even though they seem to match perfectly, because their civilian counterparts require a license that the military doesn't provide in its training. Others include nuclear power generation operations. Though commercial nuclear power very closely resembles Navy nuclear power, all Navy nuclear power veterans still have to complete the civilian licensing requirements to work in a commercial nuclear plant.

Specialty power companies are very inclined to hire power production specialists from the Air Force or prime power specialists from the Army. Gas turbine power plants also hire Navy gas turbine mechanics or electricians for their plants. Companies that do electronic test equipment calibration specifically hire veterans who have had the precision measurement electronics laboratory (PMEL) technician course that's taught at Keesler Air Force Base or, in the Navy, the general-purpose electronic test equipment (GPETE) specialty.

Construction equipment mechanics are hired by companies such as Caterpillar and some of the other heavy equipment companies specifically for their experience working on that type of construction "yellow gear." All of these are very easy to figure out, because the military person is using the exact same technology and the exact same kinds of equipment that are used in the civilian space. People from these military specialties typically require little to no training to adapt to their roles as civilians, making them attractive candidates to the companies.

Of course, another place where you can have a literal qualification is if you find yourself doing your military job for the government in civil service or as a government contractor. Most of the employers for these types of jobs typically won't take military candidates from outside of the exact or closely related specialty, because those skills are found in candidates coming from active duty. If they do choose to take someone from outside of the specialty, they are usually paid at a lower rate if they have to be trained at all.

The more common way for the service member to qualify for a civilian job is through skill sets that relate to the functional duties they've had and the soft skills they've developed that can be applied to almost any workspace. What is both interesting and encouraging is that most of America work in relatable skill-fit jobs and use relatable skills as a way to change industries or move to different types of jobs.

A close friend of mine was a military police officer in the Army on active duty, but he used his military leadership experience to move into a manufacturing leadership role when he got out in 2007. Fourteen years later, my friend accepted a position as president of a $400 million division of a publicly traded company, leading more than 1,200 people in multiple countries around the globe. My friend's career demonstrates the power of using relatable skills to change to another industry and then building and leveraging literal and relatable skills to power a career that reaches the highest levels in the business world.

Every organization is unique and requires some type of training or orientation for people to do the job. Ultimately, the key to relatable skills is: Can you learn something new? Will you learn something new? And do you actually want to learn something new?

This starts with your military training. After you finished boot camp or your officer indoctrination program, you were trained by the service to do a job and then had to apply that training in the real world of the armed forces, and any training that you've been required to use in your job is valuable to you in making your case to be trained quickly. As part of your resume, application, and interview, you will want to articulate the details of your training, how you applied the training to your military job, and then how you applied the training and experience you had to another part of your military job where you had no training. This not only demonstrates your ability to learn but also to solve problems. This experience extends into the functional duties you performed and enjoyed most.

Next, where did you get the training that allowed you to perform your functional duties? Is that something you learned in a training course? Did you learn it on the job? Did you self-teach yourself how to do that? It's good for you to understand where that comes from.

Let's look at some of the typical duties an infantryman does every day at the E-4 or E-5 level. First, you meet with their supervisor to get the plan for the day. Then, you meet with a team to review the plan and make the assignments. You address the key priorities in terms of the operations that you have to plan: coordinating with the other team, preparing your people for the operation, and ensuring that they've got equipment and supplies. Then you lead the operation, whatever it is that you've set out to do for the organizational goals. You also review your progress.

You have to make sure that people are accountable for what you've given them, so you check up on them. You have to make adjustments to your plans. You have to reallocate resources or shift priorities. You then have to report status to supervisors. Next, you complete the necessary paperwork. Then, you train your people as they need it or get them scheduled for training. You encourage your team and lead them by being there and checking in with them.

Of course, you're experienced in sanitation, organization, and staging, as all military members are. Even in infantry, there's an element of maintenance. At some point, you have mechanical maintenance duty, where you have field-stripped a rifle and put it back together. Beyond that, you're experienced with following standard operating procedures. These are all important, relatable skills that employers are interested in.

It's not a given, for example, that a person who is a civilian has ever followed standard operating procedures, much less trained other people to follow them. It's also not a given that the average civilian understands what it's like to do proper sanitation, organize work, or stage things properly so that they look sharp. These are all things that you've experienced from your military background that any active duty person could apply.

The next phase of this process is to take these things that are so ingrained in what it is that you do that you probably don't think about them—and you probably don't value them much because you do them every day as a matter of rote—and look for them in the job descriptions that you read. Don't focus on the specifics but on the basic context of a particular type of duty.

After you have looked at these, the next thing to demonstrate for a relatable skills job is how you take your training and experience and use them to solve challenges in other areas. For example, the infantryman we just spoke of was trained in how to discipline himself and follow standard operating procedures. If he takes that information and experience and uses it to lead other people and train them to be successful, that's a valuable skill.

Why would employers care? The importance has a great deal to do with how you approach your work and what people's long-term capabilities are. Mike Tomlin, head coach of the National Football League's Pittsburgh Steelers, said it best: "The more you can do, the more you can do." Like Coach Tomlin, every team leader likes having people who are capable of doing more, because the extra value builds the capability and capacity of the organization.

How does that apply to you? Well, the more value you bring to the organization, the more opportunity there is for you to earn and have a productive, happy career. So, these are the key points. Inventorying your training, understanding how you have applied that, and knowing how you have taken that training and experience to do something at the next level of your career demonstrates your potential in using literal and relatable qualifications to create success.

The Challenges

So, what are the challenges with this? Well, as we alluded to before, the first thing we come in contact with is that military and government language is frequently not the same as job description language. To say that it's like a foreign language is fairly accurate. Of course, you're not surprised to hear this because that's probably why you're reviewing job descriptions—to see how your functional duties align with the requirements of the job you would like to apply to.

Now, every one of these jobs has some type of context to it. For example, you may read the description of what a plant manager does and say to yourself, *I can do all that, too, so should I apply for a manufacturing plant manager position even though I've never worked in manufacturing?*

Well, there is a saying in the airline industry that 98% of the things an airline pilot can do can be taught to an 18-year-old kid coming out of high school with a minimal amount of training. The problem is that it's the 2% of the things an airline pilot has to do when they really need it that requires a certain degree of expertise. And the same principle applies when we're trying to determine how far we can stretch into a new position. An employer is going to look at the potential for failure in bringing somebody in from the outside who has relatable skills and can be trained to do certain things but is not an expert. Therefore, the magnitude of the responsibility of the role and the implications of failure in it will determine how far a person can stretch relatable skills in a single move.

On the positive side, I will tell you that the limits of your relatable skills are likely not as strict or restrictive as you may think. Certainly, when we talk about aviation, the fear of airplanes falling out of the sky with hundreds of people on board is a big deal, and so is running a multimillion-dollar manufacturing operation for profit. But for most of the work that veterans in transition seek, opportunities abound in roles where experiences and training gained from military service relate very well to appropriate levels of responsibility and have been highly correlated with success.

If the learning requirements to be successful in the job are not that extensive, military members coming off of active duty typically have a distinct advantage. The military environment requires constant change and adaptation, which cultivates a mind ready for new training and application. As we look at any new job, we know that we are going to have to learn some things as part of joining the organization. The question is: How fast can a person joining the organization reach an acceptable level of competency? We evaluate this every day in the search and placement work we do as part of our Absolutely American business unit.

We place a lot of veterans with maintenance skills into field service engineering—military technicians who have been trained in electrical and mechanical technologies for ships, aircraft, submarines, vehicles, tanks—you name it. Leaving active duty, most are not going out to work on military equipment but some type of machine used in a commercial industry. In the case of Absolutely American, this is typically in high-tech manufacturing, robotics, CNC machines, industrial lasers, energy production/transmission, and medical applications. The employers who use our services to hire veterans from active duty understand that they will have to teach the people they hire the specifics of the machines they make or maintain, and many of them have remarked that their veterans complete their training ahead of their nonmilitary hires—often in as little as half the time of their non-military counterparts.

This is the general view of most employers when it comes to hiring for field service: "We're the experts on our machines, and unless you've actually worked for us, there's no way that you'd ever be exposed to our machine. So, we understand that we need to teach you our machine, but from what we need in the short run, here's what we can't teach you: We don't have time to teach you electronics because electronics don't behave any differently in our company than they do in the military. We don't have time to teach you mechanics. A crescent wrench is still a crescent wrench, regardless of where it is that you work. We can't teach you safety. We can't teach you respect for authority. We can't teach you goal orientation. We can't teach you communication skills and the desire to do a good job. These are all things that we can't teach you. You have to have these things when you show up. The one thing we can teach you is our stuff, especially if you have shown us that you have been trained on how to work on a piece of equipment and understand the electronic, electrical, and mechanical aspects of technology. Show us that you have had to apply your military training on the equipment you were trained to maintain and that you used that training and experience to figure out how to solve problems with other pieces of equipment with which you have had no prior familiarity."

In this case, what we see is that the risk to the company is this very narrow sliver of the job qualifications. So, the ability to take training, apply it in your military work, and use that training and experience to solve new problems creates a strong value proposition for any military member leaving active duty. It should be no surprise to learn that institutions of higher learning coveted veterans as students even before the benefits of the post-9/11 GI Bill, which made pursuing veterans highly profitable for schools.

The bottom line: Establishing your ability to be trained, willingness to be trained, and excitement about being trained raises an employer's confidence that you will be a successful hire for them. The same can be said for applications to degree programs and franchising agreements where franchise owners have to be trained on business systems.

The second challenge: If you're looking for a leadership or management position in the civilian world, there are some unique aspects of leadership and management in the military that are very different. First of all, the military leadership and management structure is designed for combat survivability and not necessarily for cost. This is a big deal.

What do I mean by that? Most of our officers and senior non-commissioned officers will talk about the number of people they were responsible for. A division officer in the Navy can easily have anywhere between 15 and 45 people working for them. A senior NCO, E-9, could have hundreds of people that they are technically responsible for. The difference is that the military has so many levels of leadership that the actual accountability for that group is spread over a wide number of people.

For example, take a Navy division officer working on a ship. The officer is going to have a leading chief petty officer who is not only there to lead the division but also to train the junior officer. That leading chief petty officer is going to have at least one leading petty officer who is likely an E-6. The leading petty officer is going to have somewhere between three and eight people who are part of one of his work centers. Each of those work centers will have a supervisor who will lead one to three junior petty officers who are responsible for leading and teaching the new recruits how to do the job. So, with 15 to 40 people in a division, there are literally four layers of leaders, from the division officer down to the person who performs the work of the division.

As civilians, leadership and management are much leaner than that. In the civilian environment, that junior military officer who has 15 to 40 people will typically have all of them reporting directly to him or her. There will be nobody in between them. Sometimes, you may have the benefit of a lead person who helps, but ultimately, it's a single person responsible for a small group. So, when a civilian takes responsibility for a particular organization and that organization fails, the responsibility for that leadership and that failure lies on one person and not a series of people. That is one of the things that military members have to come to grips with on the outside. The number of people they have to work with is one of the reasons why many veterans choose not to pursue that.

Now, here's the positive side of this. When you are working with civilian employees, they are not your responsibility 24-7-365, like your subordinates are in the military. Civilian employees are only your responsibility when you are on duty. That doesn't mean that the operations and what you're responsible for don't go home with you, but having to bail somebody out of jail for an incident that happened outside of work will not be something you will have to deal with as a civilian.

Another aspect of leadership and management unique to the military environment is that no matter how we cut it, the military has a rigid rank structure and military culture that is backed up by the Uniform Code of Military Justice. In the Introduction, we talked about how you can be thrown in jail for missing work. If you do not do what your boss tells you to do, you can have a significant reduction in your pay, have your liberties restricted, and even be put in the brig. You can also have your rank reduced permanently as part of getting "busted" in a non-judicial punishment. Getting yelled at is an expected part of military culture. The current leadership tends to discourage this approach, but it's still in movies and an expectation of American culture as a way to develop toughness—especially in boot camp. In the civilian world, it's not, and it will get you fired.

Also, there is a rigidity to the way that the military works. You almost never see E-6s working for E-5s, and the difference between them often comes down to nothing more than age and time in service. That rigidity and the entire military culture can sometimes make it a little too easy for military members to shortcut good leadership and management for the short-term results that can be had with a form of rank intimidation. Civilians have different rights and expectations than we do in the military, and this can sometimes be challenging to the veteran coming from active duty.

The third challenge has to do with employee demographics. The military is largely a young person's business. More than 98% of military members are below the age of 45, and very few serve beyond that age. As a matter of fact, we use the term "retired" for somebody who could be as young as 37 to 38 years old. In the civilian world, the term "retired" typically causes people to envision a person who is at least in their 60s and has already lived most of their productive life.

So, the age range of employees that you see in the civilian space is going to be much broader than what it is that you see on active duty. You are also going to see people with more varied backgrounds. When compared to the consistency of how the military recruits with standardized testing, trains using the same types of courses, and employs in a consistent environment, the variety of backgrounds found in a civilian organization can make evaluating talent difficult for the veteran. At the same time, you will no longer experience the same kind of national diversity but mostly the demographics found in the areas where you live and work.

All those things can be difficult for leadership and management. What we find in the transition with relatable skills is the lack of exposure that the military member has outside of the armed forces or the federal government. This can arise when dealing with direct sources. For example, in the case of purchasing, many military members who are involved in supply and acquisition don't deal with direct sources. Instead, they deal with the military stock system. Though there are a lot of similarities between the two, there are also a lot of differences. The federal government and the armed forces have a lot more guardrails in terms of what can and can't happen using stock than the civilian marketplace.

Fourth is budget control. In the military, we play with "funny money." We don't know where the money comes from to pay for everything. There's no connection between the budgets we have and the people on our staff. We have no exposure to how these people are hired, trained, and developed before it is that they come to us in our units, and we have no control over these types of things. Therefore, one of the biggest things that we have to understand in coming out of the military and going into the commercial space is that civilians generally know and understand that it takes money coming from someplace to pay for everything, including their salaries.

This is important because it connects to the job search process. When employers look at you, in the backs of their minds, they are asking themselves, "Why should I pay you at all? Why should I trust you and obligate myself to give you a salary? And if I do pay you, how much is what you can do worth to me?"

It's said that cash flow is the lifeblood of business. If you have no cash flow, you have no blood flowing through your system, and you die. So, the entry of cash into a company and how it's used to develop a valuable result that can be sold is of vital importance to civilian companies.

Consider this common misconception among senior officers: Military officers are never CEO equivalents. Why? No military officers ever have responsibility for sales and revenue generation as part of their operational or administrative responsibilities, which is one of the most critical areas of expertise for CEOs. Therefore, any military officer seeking a role that includes responsibility for sales and revenue generation needs to understand this gap in experience and its significance in any nongovernmental organization.

Finally, F-bombs and other foul language issues can be a big problem for some. As much as I hate to say it, we sometimes use F-bombs in the military the way that many people say, "Um." And though that is largely still pretty widely accepted among active duty service members, that type of language in the civilian world is not. It's a part of military culture that has gotten many veterans into trouble. A good rule of thumb is to always be your professional best in the workplace. Regardless of your military experiences, all veterans know what this means and should govern themselves appropriately. No excuses. The world is watching.

It's been my experience that these are the five most frequent challenges that veterans face when entering the civilian work environment. According to government statistics, only 6.1% of U.S. citizens are veterans, and as of 2022, only 8.5% of the 6.1% of veterans in the U.S. population were under 35. That means that only 0.5% (one out of every 200) of the U.S. population under 35 is a veteran. So, we need to understand that people don't speak our language, and because we are going into their world, we bear the responsibility of learning their language and their environment more than they do ours.

Why? Because our failure to do so is going to hurt us more. When you decide to buy something, you accept accountability for it. Your prospective employer will be thinking the same way, and your employer will not accept the responsibility of paying for your services if they cannot understand what you can do or have concerns about how you will behave.

Again, the good news is that employers tend to value military experience highly. Among other things, this is due to your personal character qualities, respect for authority, goal orientation, ability to get things done, willingness to work hard, and ability to work with diverse populations. As long as you can inspire confidence in your ability to reach acceptable proficiency within three months without being a headache for your employers, most of them will be willing to give you a chance to prove yourself in their environment.

We've all heard the phrase, "Perception is reality." Nowhere is this more true than in the job candidate evaluation process. Regardless of your actual skills or how little difference there is between your military work and the job you want, the employer's perception of the gap between your skills will impact their decision to hire you or not and what to pay you if they do. Small gaps lead to high confidence, which leads to offers and highly competitive pay rates.

So, as you're looking at roles to apply to and you're evaluating them against what you have, look at the requirements of the role. Ask yourself, *Is there a big risk to the employer that I may not be able to get this within the period of time that they would need me to become proficient at the role?*

That's the timeline that you should be looking at in determining how long it would take for you to develop proficiency in what they were doing. We'll address this further when we get into the interview itself. Here's the caveat. Remember that as you're looking at jobs, most military members will tend to devalue the experience they have from active duty versus the civilian experience in that community.

If you are in doubt about whether you may be qualified, I encourage you to apply because you're probably devaluing yourself and overvaluing what they have. Try to focus on identifying what the employer needs the person they are looking for to accomplish, and be sure to address that as part of your application.

Applying for Jobs

So, moving on to the applications themselves, what are the different ways that you can apply for jobs? Most of these are obvious in today's world. Everybody has heard of Indeed. Everybody understands that companies have websites. Everybody understands the nature of going to a job fair and meeting with people. These are all ways that you can apply to a company.

The truth is, in today's world, nearly all applications are done online. This change has happened in just the past 15 years. In the early 2000s, you could still apply to a company by bringing in a resume and handing it to them. Now, companies of any considerable size are almost all doing their applications online. Most smaller companies will still be less formal with this because they don't have the same kind of requirements.

However, companies that are over 50 people, as a result of a number of laws that have been enacted over the last 20 years, are now required to do some form of online application to be able to track their statistics and defend their hiring policies. Consequently, applications are ultimately going to lead to something that is done online, either on a company website or a job board.

What is included in a basic application can be as variable as the imagination. The application is minimally going to require a resume, which includes what you have been responsible for and what you've done with those responsibilities. The employer will be looking at the relevance of the functional part of the job and your proficiency in that, combined with the responsibilities you've had and how you fulfilled them relative to your employer's expectations. Then there are usually some types of questions related to your reasons for being interested in the role and your beliefs about your qualifications related to it. On most applications, there is usually a place for comments or an opportunity to include some type of cover letter, statement of interest, or description of your objectives.

Though your history and qualifications are fixed at the time of your application, your motivations are not. As a company goes through a continuum of evaluation, there is no substitute for desire. Since 2012, we have observed an increasing trend among employers, and now we find that most prefer candidates whose motivations, interests, and value systems align with the company and role first.

As long as the candidate can meet the minimum level of proficiency, it is no longer the most qualified candidate who wins the day, but the one the employer knows will get energy from performing the duties of the role and fits the culture. This is because this person will require little to no supervision, produce high-quality work, and remain with the company longer than others. People want to work with those who like what they do.

Employers now take a much more long-term view. A few months of ramp-up is a small price to pay for a top-performing employee who stays with the company for a long time, and if you have upside potential for more responsible roles, even better!

We've all worked with people who didn't enjoy their job, and we all know how those people suck the life out of a room. On the other hand, we've all worked with people who love the work that they do, and we know how that encourages other people to love what they do, too, and creates a much more enjoyable work environment. People are not any different at their core level when it comes to the civilian world.

People want to work with people they like, and the people they like are typically people who are enthusiastic about what it is that they do. So when you're applying, this is your opportunity not only to tell a company that you are interested in them but to tell them **why**. And the connection with this is so important to **make the connection with the problem they are trying to solve.**

Remember that in all cases, the sign they post says *"Help Wanted."*

That's what they're looking for. So, if your introduction focuses on what you believe they need to solve that problem, you will get their attention. Sprinkle in your enthusiasm for that work, and the likelihood of success will be much higher for you. If you can grab phrases from their job description or their website that they use to describe who they are and what they look for in people, you are going to be speaking their language. That is how we influence people.

Think about a time when you thought you didn't like somebody and, in an unexpected moment, that person offered to help you with something. What happened to your feelings about that person? Of course, they changed in an instant! People like people who care enough to help them. It's a natural human instinct.

If you want to help me with something important to me, I already like you better. The opposite is also true, just as we discussed in the networking section. If the only thing that you are interested in is what's in it for you, it turns people off.

We see this mistake at job fairs all the time. A military member will walk up to an employer and say, "I'm a (service), (rank), (specialty) with (number) years of service. What do you have for me?" The gut-level response from every employer, which they don't say out loud, is, "I don't have anything for you, you lazy, self-serving piece of garbage." It is the biggest turn-off in all of military recruiting, yet it is repeated over and over again.

No, the burden is on you to tell me what problem you believe you can help me solve. If you are coming to me with a solution to the problem that I may have, then you're getting my attention. This is not new to military culture. Your senior officers have told you multiple times, "Don't bring me a problem without bringing me a proposed solution." Why should we think that civilians are any different? And why do this with an employer that you need to like you so you can get paid?

You have a problem. You need a job. You have a day when your pay is going to stop. That's a personal problem for you. They have personal problems, too. They need people who can solve the problems that their businesses are facing. It is possible that the two of you may come together to find a mutually agreeable relationship where you can solve their problems, and they can solve yours. But the onus remains on the job-seeker to make that introduction and address that. Even if you don't know the company and what it is that they do, telling them the things you like to do and do well is at least a minimally beneficial way to approach the problem.

Of course, having the specific problem in mind is better, but let's use an example. Let's say that you have been a network administrator for your unit, and your responsibilities have been maintaining servers, user accounts, and desktop support. You approach a company at a job fair and say that you would like to acquire a similar position.

Your proposition of a solution you could provide might sound like this: "**It's been my experience** that not all network administrators are created equal, and among the things that most affect the teams they support is responsiveness to trouble calls. I have been skilled in working with Cisco servers, desktop support, and a number of issues in the IT technology space. One of the things that I've been noted for has been giving really good support to the teams that I work with. If your company is looking for someone who can provide better IT support to your team, I would be interested in speaking with you about the role and discussing how I can do that for you." Simple! And you don't even need to know anything about the employer. You just have to know your job and how you have exceeded expectations in the roles you've had.

The idea behind this in sales is a concept called "features and benefits." If you've been through the military recruiters' sales training, you've heard this. Features describe what the product is, but the benefits are what the product does for somebody. Ultimately, we don't buy features; we buy the benefits of those features. So, your experience in working with Cisco servers and doing desktop support is a feature. In the job world, we could call this a "qualification." You have this particular qualification, but it's the benefit that helps me relieve pain or enjoy pleasure that gets my attention. In this example, it is quickly relieving the frustration of having to wait for technical assistance when the technology that allows an employee to do their job fails.

The person in this example may add, "And by the way, I'm very good in front of customers because I understand respect for authority and goal orientation. This is what has helped me to be successful in the military, and I'm looking for an organization that values the same types of things." Those are the kinds of statements that get people interested, and when you can not only tell them what you do but how you have made an impact through the work you've done, you are solving a problem for someone.

We have all had something where we have been able to use our talents to deliver a result that is important to our employer, and including things like these in your application, resume, cover letter, or personal introduction will get employers interested in you. This is why these employers are investing time into this process in the first place. They have a problem that demands a "people solution" for their business. When you show them that you are in the business of solving problems for other people using a particular set of skills, people will listen to you.

Now, who are the people that you're going to talk to as part of the application process? Most frequently, it's going to be somebody from either human resources or talent acquisition. There are times when you may have the opportunity to talk with the manager directly or with an executive. However, most people, especially if they're coming from the military, are going to find their applications channeled through talent acquisition and human resources more than anyplace else.

Here is what you need to understand about engaging people in human resources: such people have very, very broad responsibilities. Their role is to help managers and executives be more effective with their time by casting out nets in the form of job descriptions and participating in different types of activities that help them find candidates for the jobs they have, and the goal is to attract as many qualified candidates as they can. The problem is, just like dragging a net to catch fish, you're going to catch some that you want and a number of them that you *don't* want.

The human resources and talent acquisition professional is the person whose role is designed to separate those that they believe the company would be interested in for the role from those they would not. This is an important part of the process. **This person is not there to help you get a job.** Though they have an outreach function going into the market, that function has a very specific *responsibility*: find the people the organization needs for the jobs it has. That is their first and only major responsibility.

To make this person's job as easy as possible, help them to see that you fit what they are currently looking for. This is why research is so important: It allows you to understand what the need is and how you can help solve that problem. Then, you either apply to them through a website, talk to them in person at an event, or send an introductory email with your resume. This is what you have to focus on.

You can be certain that this person sees lots of resumes and applications every day. As a matter of fact, it's been stated that the average time that an HR or talent acquisition professional places on a resume or application that comes into their company is a mere eight seconds. If you haven't established that you're bringing something to the table that the company needs within an eight-second glance at your resume, then your application is going to be put onto another pile, and somebody else who has made that job easier for them is going to go forward while your application is left behind.

Second, because of the volume that HR and talent acquisition people see in applications and resumes, savvy job-seekers need to understand that once they've viewed your resume once, they aren't likely going back to it again. Internal recruiters and talent acquisition professionals are not known for going back to their own databases to find people they have already reviewed and eliminated from consideration for a position. When a new job requisition is generated, their normal approach is to look at the new applications for that role rather than go back and look at old ones. The general belief is that the applications in the previous job are stale. It's only whenever they are not getting acceptable and sufficient new applications from their postings that they will even consider going back to old applications.

The only way to be sure that your application is fresh and reviewed by a talent acquisition professional is to make sure that you apply for the job you want.

The second person you will encounter in the application process is the manager who is responsible for the open position, the "hiring manager." **This person is your primary customer.** The talent acquisition person serves this same customer, and that is why you have to appeal to talent acquisition and the hiring manager in most cases, because talent acquisition has to approve you to be reviewed by the hiring manager.

Talent acquisition professionals help hiring managers solve the problem of finding the people the managers need while they are busy doing their jobs every day. These managers don't have the time or bandwidth to evaluate every single candidate that comes into the company. They need help taking a large list of applicants and shrinking it down to a smaller one, from which they can then select candidates for interviews.

Compared with the talent acquisition professional, the hiring manager's focus is personal and real. The talent acquisition professional's responsibility is primarily built toward creating candidate flow for the managers who are looking, and that's typically how they're evaluated. Do they bring in enough candidates for the manager to evaluate who are properly screened and can ultimately be hired for the position?

But for the hiring manager, the pain that this absence creates is personal. They may be having to overwork their people or themselves, or they may be getting pressure from their manager to fill the job due to a growth or operational goal. There can be any number of reasons why hiring managers feel pressure to fill a role, but it is that pressure that they most want to relieve—as long as they don't create more issues for themselves, including having to go through the process again in the short term. The savvy job-seeker understands this and uses it to their advantage, something we'll discuss more when we cover being interviewed.

The third level of person that you can speak with as part of the job search process is the company's executives. The executives, or senior managers, are responsible for the success of the company in their area. These people care about the team, about who joins it, and also about having a connection with the people who are hired.

Their role in the hiring process is typically as an additional check to make final approval on hiring somebody, and they frequently keep their eyes open for things that are related to the candidate's fit. Ultimately, the role of the executive tends to be very broad and forward-looking. Most executives are happiest when they're solving tomorrow's problems, not today's. As a general rule, we are talking about the person who is at least two management levels above the job you are trying to get.

Understanding the Executive

The happiest executives want to be able to do their work and plan as far down the road as possible. A big part of the job of an executive is guiding an organization into the future, and that is where they most want to apply their talents and abilities. Executives don't want to deal with today's problems.

So, when you're speaking to executives, discuss how you can help them get an immediate return on your being hired and how your presence there can add long-term value to the company. Long-term employment, happy people, and the ability to take higher steps in the organization are all ways you can add long-term value.

Chapter 6
The Interview

S ometimes, the easiest way to win the job you want and get the best offer is to excel in places where it's not expected.

Interviewing has gone through significant changes in recent years. Just a few years ago, very few companies were comfortable hiring a person without an in-person interview where the manager had the chance to size a person up with a handshake and look into their eyes to see what was really there, but the COVID-19 pandemic changed all of that. The inability to meet people in person necessitated finding new ways to make decisions, and the rise of online meeting technology helped solve that problem. It provided all the face-to-face experience you get in person without the physical contact. Whether you consider that good or not is a matter of personal preference, but now it's very common for employers to make hiring decisions without meeting onsite, even though we have been out of the pandemic for a number of years. I have done that myself on a number of occasions.

In nearly all cases, an interview with an employer is conducted in one of three ways: telephone, online meeting (with or without video), or in person, and these will all be discussed as we look at the phases of the interview process.

The Initial Interview

At this point in the process, you have gotten clarity on the functional duties you have performed that you like and want to continue to do and the functional duties you have not been able to do but would like to add to your job description. You are confident that you would be able to become at least minimally proficient in the functional duties and key responsibilities you want to add within 90 days, so your hire poses minimal risk to an employer.

You have also found a number of job postings whose descriptions include these duties as you have them written or in slightly different wording, and ideally, you have engaged some transition assistance professionals to help guide you and maybe help provide you with opportunities. In some way, you have made an application with these organizations that match the duties and responsibilities you want and fit within the parameters of your transition wish list. You know your well-considered application has gotten their attention because they have indicated to you directly or through your guide that they want to set up some kind of conversation or interview with you.

Now this conversation has been scheduled and is about to happen, and the initial call is on the telephone. As we discussed in the previous chapter, the first interview is typically with someone in human resources, but depending on the company, it could be with a manager. Almost always, however, the purpose of the telephone interview is an initial screening.

Initial interviews are not usually intended to go into a great deal of depth but to validate the information on your resume, your personal information, your interpersonal skills, and your interest in the role. All of this is used to determine if it is worthwhile to proceed with a deeper interview that might be done either on the phone again, in person, or perhaps using video meeting technology. What is most important for the savvy interviewee to know is that **First-round interviews are not screening-*in* interviews; they are screening-*out* interviews.**

The objective is to determine whether your application is worth pursuing or not. While there may be a slight tendency to screen in candidates, especially when the pressure to find talent is high, this process is generally designed to narrow a larger pool of applicants into a smaller, more qualified group.

This is why it's very important to remember that with initial interviews, the business problem that is causing pain for the hiring manager is the only reason that they are in front of you and why they posted the job in the first place.

Addressing that issue while ensuring that you don't cause any yellow or red flags to pop up in the mind of your interviewer is going to be the key to getting to the next step.

So, how does this happen in a telephone interview? Well, the first key is always preparation. This preparation goes beyond the research you've done in making your application. It is the preparation to present yourself most effectively. If you remember from the previous chapter, you will be most effective in influencing people to hire you when they feel that you can quickly alleviate their pain without creating any downside.

That preparation starts with getting yourself into a place where you can control the environment around you. You don't want any distractions for you or the interviewer, so interview in a place where you can control the sound levels.

Second, most people have a degree of discomfort with communicating on the telephone because they rely heavily on visible feedback from new people they meet to determine if they're connecting well. The telephone takes that away and causes us to have to focus only on the words and sound of somebody's voice, which can be a little bit intimidating to people who are not primarily audible learners.

However, techniques can be used to set ourselves apart from other people in this process. The first among these is standing up during a telephone interview. Standing is the best way to project more enthusiasm through the telephone, and enthusiasm for the role and for meeting a new person covers a multitude of sins.

Just the simple act of standing relieves pressure from your diaphragm and allows you to enunciate words and project your voice more clearly than when sitting. Your ability to communicate clearly and enthusiastically is an important component of getting the person on the other end of the line excited about you. Again, people want to work with people they like, and they tend to like people who like what they do. For the best results in a phone interview, stand up!

If you have a headset for your phone, that is even better. The freedom of movement that comes from standing and being able to freely move your hands will help you express yourself verbally the same way you would in person, and that will project through the phone. We call this "embracing your inner Italian." If you've ever been to Italy or any Mediterranean country, you will find the people to be very passionate and expressive. You will frequently see them moving their hands while talking on the phone, the same way they would when they are making an emphatic point in person. Their emotions cause them to move their bodies and arms. Freeing your body to move will allow your positive emotions to flow throughout your body and into your voice.

Our bodies are connected by a series of neurons that work in different ways all the time. A perfect example of this is our emotions. What we're talking about here is that our *motions* can change our *emotions*. If you don't believe me on this, think back to when you were a child and down in the dumps. When someone told you to "Smile! Stand up straight, and put your shoulders back!"—and you did—you stopped feeling down in the dumps. You don't know why it worked; you just know that it did.

The very act of doing something physical—smiling—causes the neurons that connect to happiness in your brain to engage because the neurons that were used to move your muscles in this direction were moving toward a happy place. The brain says, "Well, I'm smiling, so I must be happy." So, that's what you are: happy.

The same principle applies when you're talking on the telephone during an interview. If I were in person with somebody and I were talking, I would be leaning forward, using my hands, and moving, maybe even walking. So, if you're doing a phone interview and expressing things enthusiastically, move your body because that will allow you to project the same energy as if you were in person.

The next item in your preparation is figuring out how to dress. This may seem counterintuitive for an interview where the other person won't see you, but your dress has nothing to do with the other person. It has to do with you. For many people, being their best requires dressing the way they do when they are at their best. This is different for everyone, and you know you. Therefore, we suggest that you prepare for your phone interview by planning ahead on how you will dress to give you the comfort and confidence to be successful. Some people perform best when dressed formally. Others like to be more casual. Whatever makes you feel confident in telling your story and meeting someone new is the right thing for you.

First Impressions Matter

We have all heard the maxim: "First impressions are lasting impressions." When coming in contact with people for the first time who will also make financial decisions about you, make sure that your image reflects the kind you want to be connected to, because whatever makes the first impression with them will last. This process started with your resume and application, but there is more to consider than just these.

Having an email address that sounds professional is important. Typically, in today's world, most people will set up a Gmail, iCloud, or Outlook address. We recommend using one that uses your first name and your last name. People will often include a number or something like that to make sure it's a unique identifier. This helps keep it professional.

Second has to do with your telephone. Most younger people today don't use voicemail much, if at all, but the people who are contacting you for phone interviews are likely to leave a voice message for you if you are unable or unwilling to answer your phone. Therefore, the message that goes out on your voicemail system becomes an important part of those first impressions you make on employers. Make sure yours promotes you at your professional best. If you're not comfortable creating a friendly and engaging voicemail message, at least make sure your voicemail is set up and uses the standard message provided by your phone carrier.

Also, be aware that many wireless phone carriers now have separate voicemail boxes for people you don't know. T-Mobile, in particular, has recently added this to their phones, and the way they have you opt into the service may cause you to miss important messages from employers. So, check that out on your phone and with your carrier to make sure you receive those messages.

As for the things you broadcast to this world, I always remind people: If it's not something that you would be comfortable saying to your grandmother or face-to-face to the captain of your ship or the commanding officer of your organization, it probably shouldn't be broadcast on anything.

And that leads us to social media. Your content on social media is not as locked down as you may think. Your comments, pictures, likes, messages, and posts are on the internet. If your brand on social media does not match the image an employer wants for its people, your interview may go well, but you may find yourself waiting for a follow-up call that never comes.

As for the phone interview itself, you will need to be prepared to discuss your qualifications in relation to the job. Your education, training, and experience are not a biography. They are the basis of your expertise to perform the job at hand. When you tell your story, everything you say needs to relate to what the employer is seeking in the role, and regardless of the role, the ability to take training and apply it and the ability to adapt to new environments are strengths of your military service that are always applicable. Be ready to describe the most relevant of these in detail to demonstrate their depth.

You can be confident that the person interviewing you wants to hear specific details from your background that are relevant. So, in getting ready to roll out your qualifications, prepare yourself to describe your past education, work, and training in detail but with brevity. Your education, training, and experience form the basis of your expertise, and the specific details you provide will help the employer assess the gap between their job requirements and your qualifications. Details of interest to an employer include not only what you did but how you did it. Describing the steps you took to solve a problem or overcome a challenge usually requires words and phrases that are easily understood by anyone. My advice: *Practice describing what you have done as if you were telling it to your 15-year-old niece or nephew. If you can describe your work so a person at that age can understand it, you are likely to succeed in communicating your work to an interviewer who doesn't understand the military lingo.*

Preparing Questions that Make an Impact

Next, you will need to prepare questions about the expectations for the role—particularly what will need to be done and how the employer wants it to be done. Remember that the process of selecting someone for hire is about finding someone to solve their problems. Doing that effectively means understanding the specific details of what they want the person to do and how they want the person to do it. These details matter a lot to them. So, based on what you have seen in the job description or have learned about the organization, you need to develop some questions that will help you better understand the job, the organization, and the work environment at this phase of the process. Taking these questions right out of the goals that are part of the job description is a great way to be prepared for this. In every case, your curiosity about the specifics of the expectations for the role conveys your interest in actually solving their problems, which is something they hope to see in meeting with you.

Answering the Call

When the call finally comes, professionally answering the phone will serve you well. Even today, military members are taught how to answer the telephone. Sometimes, they're told to include a lot of information and can speak that very quickly (sometimes too quickly), but we're all told to answer the phone using a sequence like the following:

> **(Greeting)** Good morning!
> **(Identify Yourself)** This is Mike.
> **(Find Out the Reason for the Call)** How may I help you?

Answering the phone this way characterizes you as very professional. It demonstrates to your interviewer how you will respond to calls at work, and it subtly conveys a message of your willingness to be helpful right out of the gate, which makes a strong first impression.

The employer will then start with some type of introduction of who they are, and you will get underway from there. Usually, the interviewer will not immediately start asking you questions. Instead, they will start by making small talk, explaining who they are in their organization and some things you may or may not know about the role or company. At some point, the conversation will turn toward questions about you and your background.

This is where you may be asked to provide your narrative: "Tell me about yourself," "Walk me through your resume," "Tell me about what you did in the Air Force," or something similar.

A Powerful Lesson

Let me illustrate using a personal story. What I am about to describe here is something that I learned and recognized as powerful from a series of presentations made to me while I was serving on a committee for a municipal main street beautification project.

About 15 years ago, my local community was in the process of using Main Street funding to improve the townscape of a street that ran from the small town into the township where I lived. We had previously sent the engineering firms in the area a request for proposals to help with that project. As a parallel to the job search process, this would be the same as a company posting a job. In reply, we got a number of proposals from firms on how they could handle the process. We evaluated them and selected three of the firms to meet with us in person to go over their proposals. What we noticed in the presentations from each of these firms was consistent across the board.

All the firms started off the proposal by talking about who they were. They talked about their education and their certifications. They talked about the experience they had as an organization, as well as the education, qualifications, certifications, and experiences directly related to the project at hand.

As I listened, I realized their approach was extremely powerful in the way that it focused the listeners on the most important qualifications, giving us the confidence that these people had been there, done that, and were capable of doing the job we wanted. They established their bona fide credentials to do the job.

This is the same thing that you're going to do. You're going to walk through your experience with a focus on the training, education, and experiences you have that line up with the ones listed in the job description or from a briefing you received from your transition assistance guide.

Sometimes, the employer will start a phone interview by telling you about the requirements of the job, especially if they are challenging, to validate your interest in the role. If you have no interest, the employer will bring the interview to a quick close. Remember that. Let me say again: One of the best ways you can show interest and set yourself apart from others is by asking questions about what the employer wants you to do and how they want it to be done. Employers love it when people are doing their best to assess the job requirements through questions. If this job is closely related to your military experience and training, you should have some insight into the challenges that come with the kind of work they want you to do. Ask about those. Find out how they have handled those challenges and share what you have done in similar situations.

The second step that happened when I was on the municipal committee was that as soon as each of those firms finished covering their qualifications, they immediately transitioned to asking us a question about something they saw in our request for a proposal that got to the heart of what we wanted to accomplish. The question came from their expertise in working on this type of project.

To use an example from my experience, we work with a lot of people on active duty who work in maintenance. At the core of maintenance is knowing how to do preventive maintenance—oil changes, grease and fittings, opening and inspecting, these types of things. They understand how to upgrade equipment, install new software, or add on a new component, and, of course, they troubleshoot and repair their systems. These are what everybody would recognize as the core functions of anyone who works in a maintenance position.

However, if you've worked in maintenance, you understand that there is a lot more to it than just the core functions. First among these is paperwork. In the military, the maintenance isn't completed until the paperwork is done. Similar things happen in the civilian world. As a matter of fact, most of the problems that employers have with maintenance don't have to do with the quality of the maintenance work, because the people they hire tend to like the core functions of the work. The problem is usually in other parts of the job, and the paperwork is usually the least favorite thing to do. Military maintenance techs know this and can converse about the importance of paperwork in comparison with the employer's expectations.

Customer service is another big component of maintenance outside of the core functions.

Whoever uses the equipment you maintain is the customer. If you work in the motor transport division and maintain Humvees, your customers are the drivers operating those vehicles in the field and the unit commanders responsible for them. So, if the vehicle performs the way that it should for the unit and allows them to complete their mission, the customer is happy. If the vehicle fails while it's out on a mission, then they are unhappy about that.

Responsiveness and communication with the customer are other functions outside of the core maintenance duties, and a military technician's experience with issues surrounding customer service can provide a good topic of conversation. Specialty tool issues and the control and coordination of maintenance with operations are also things that cause headaches for maintenance managers, and the military candidate can relate to all of them.

Enter the Expert

As soon as you finish talking about your bona fide experiences, training, and education that you have that are relevant, turning the conversation by asking the employer a question that gets to their primary objective using your experiences as a basis of the question is a powerful way to establish yourself as a strong candidate interested in the job. For maintenance, this question could be phrased as follows: "It's been my experience that making the users of my equipment happy has been the key to my organization's success. What kinds of things do you like your maintainers to do to keep the users of your equipment happy?"

Whatever the role, there is a goal you have had to reach, a people issue you have had to handle, or a work process that you have had to use that can form the basis of your question that will be common to the job you are seeking. This will be the source of your expert question, and not only will it help establish your qualifications for the job, but it will also demonstrate your interest in solving their problems. You will be off to a fast start, whether on a telephone, in a virtual meeting, or in person.

If this doesn't sound applicable to you because you don't feel like an expert on what your prospective employer does, remember that you are an expert on the functional duties you have performed and the responsibilities that you have had. Your interviewer believes that your expertise has some relevance to the work they need to get done. Otherwise, you would not be there. Someday, you will be an expert on your industry, but today, you are still an expert on you. Use the education, training, and experience you have gathered to try to understand their needs first, and then, you should have no trouble relating your abilities to what they need you to do.

Many military members come into interviews expecting it to be like going before a board, where you're sitting at the end of a long table with a number of powerful people who are making critical and harsh observations about your qualifications. They expect this conversation to be tense. They expect to be put on the spot. They expect to have gotcha questions asked of them. Though there are some types of interviews that are still conducted that way, the vast majority are not. Today's interviewers want you to be relaxed so they can be relaxed and get the chance to know the real you. A good interview, at its core, is just one or more people coming together to discuss whether they can solve the problem in a way that they like. It's as simple as that.

The candidate would not be there to have this discussion in the first place if there was no respect for their qualifications. The term "qualifications" is another way of saying expertise. Your qualifications are the things that you are expected to be an expert on, except in cases where specific industry expertise is required due to the nature of the position.

Most employers expect you to be an expert on you and not an

expert on them, though they do want you to have spent some time learning about them as an indication of your interest. At the core of all of this is that you know what you know, while they know what they know.

In my interview preparation, I try to help my job-seekers see themselves as expert consultants being called in to see if they can help get a project to success. I believe this encourages the right mind-set, because it is how people approach hiring for expertise, just like in my municipal project example. We were looking for people who had expertise to help us, and your prospective employer is, too.

It may surprise you to learn that a good interview should be very conversational. In most cases, except where positions are highly regulated or critical, your interview with an employer should be a comfortable exchange of answers and questions from both sides. After all, isn't this how things are going to work once you start? Avoid the feeling that an interview is like a board. Nobody is expecting you to sit in the proper position of attention, keeping your eyes straight ahead while giving rehearsed answers. No, they are looking for someone who will make coming to work every day a better experience for everyone.

From this point, a good interview or a good consultation, as we sometimes call this, involves a conversation where both parties are asking questions as they're going through the process. The interviewer asks you a question about something, and you answer the question. Then, after thinking about what it is that has just been said, you're going to toss something back to them to answer. This should feel like a very comfortable conversation.

Now, you may wonder why that would actually be the case. The employer knows that at the end of this process, for it to be successful, you have to say, "Yes." If you don't say yes, everything they have invested up to that point will have been for nothing. One of the reasons why I love working in recruiting is because we work with the only product that can make a decision about who is buying it. You have the final choice of where you go to work.

The employer has to give you the invitation to come work for them, but if you decline that invitation, they have to start over. Good employers understand the need for mutual respect, but many military members in transition see the job as a valuable and rare commodity that they're absolutely afraid to lose because of the time pressure and because of the minimal exposure that they have had to the job search process. But guess what? The employer sees it the same way. If you remember from previous chapters, there are not enough good people in this world to satisfy all the jobs that are there.

Let's get back to questions for a moment, because they are vitally important at all stages of the interview process but critical in the initial interview. Even though your interviewer may cover a lot of what you are prepared to ask, you still need to ask some questions. If you cannot think of anything specific because your list of basic questions was covered, one of my favorites is "Can you tell me more about (something they said about the job)? I'm interested to hear more about that part of the role." When you ask that in a tone that conveys interest in the role, your interviewer will love that.

Going Deeper into the Interview

So, what will happen for the rest of the interview? Well, things are going to go back and forth. You could be asked a number of different questions about your background, and you may be asked about your reaction to a hypothetical scenario. These are called "scenario-based questions." In these cases, your interviewer describes a scenario that could happen in the scope of your job and asks you to describe what you would do if that scenario were to unfold in front of you.

Situational Questions

You may also be asked to tell a story about something specific you encountered in the past and how you handled it. These are called "situational questions." You could also be asked questions designed to reveal your attitude on certain topics, and if the role has a technical component, you could be asked technical questions.

What is important to remember is that every one of these questions needs to have a happy ending that the employer understands. If you are asked about a time when you encountered a problem, they want to see that you successfully navigated the situation, overcame the challenge, and delivered success for the organization. If the situation involves working relationships, they want to see that you put the company's goals first and accomplished the mission but also that the experience helped you better understand the other person and improved the relationship.

If the scenario involves a potential conflict with a customer, they want to know that you can de-escalate the tension, demonstrate care for the customer, inspire confidence, and complete the work to the customer's satisfaction.

Maybe it's a leadership scenario in dealing with an underperforming team member. In that case, they want to know that you recognized the problem, met with the person to better understand why there was a problem, built a plan to get the employee back on track, provided guidelines, monitored progress, and restored the performance. (I know some military leaders may be proud to end a story like this with "and then we got him booted," but this is not the happy ending that civilians want to hear unless they specifically ask for a time you had to terminate someone.)

In other words, once again, everything I needed to know about success as a civilian, I learned in the military. These are all things that are part of our military value system.

Handling Technical Questions

What about technical questions? Even if a question about an industry-specific piece of equipment, such as programmable logic controllers (PLCs), is asked, the job description will have that requirement in it. Doing research on that piece of equipment prior to the interview will give you the opportunity to see the similarities between the control systems used in your military specialty and the industry equipment. Using the common elements between systems, a great attitude, and confidence in the ability to learn in a short period of time, you can help the employer see a happy ending to problems with equipment and technical challenges.

Showing Interest With Questions

Lastly, most phone interviews are short—typically 30 to 45 minutes, with some as few as 15 or 20. Success in an initial telephone interview means getting invited to take the next step. You are trying to get past the initial screening interview and be considered somebody worth moving forward. Sometimes, these interviews are structured to help the employer's staff, and you cannot have the kind of conversation we ideally want, but even in these cases, the end of the interview usually provides an opportunity to ask questions. Be sure to take it, and make sure your questions are focused on what the employer wants you to do or accomplish and how they want it to be done.

Imagine that you are there to troubleshoot their problem to offer a solution. When you get the chance to ask questions, try to understand their challenge and what they want success to look like. Showing interest through asking good questions about the job and expressing enthusiasm for it has turned more borderline interviews into yeses than anything else that I know. And nothing says more about your level of interest than the questions that you ask.

Every person in the world who teaches interviewing will tell you that you have to express interest in the position in the form of either asking for the job or asking for the next step at every interview. However, the best way to influence somebody to believe that you are actually interested in the job is to demonstrate that interest by asking questions that get the interviewer to envision you in the role and performing at a high level. Here are some very straightforward questions to demonstrate curiosity and interest in the role and performing it well:

- "What do you need me to do here?"
- "How can I make you happy that you hired me?"
- "What do we have to do to make sure that we win?
- "What can I do to help our team stand out in a positive way?"
- "What do I need to do in the next 90 days to validate that you hired the right person?"

Questions like these are music to the ears of the person you're talking to, because the vast majority of people will ask about something self-serving, if anything at all:

- "How much does the job pay?"
- "How much time off do you get?"
- "When is the first pay raise?"

These questions are important, for sure, and there will be plenty of time to get the answers, but they aren't going to help you get to the next step in the process and beat your competition. However, when you're asking questions that focus on addressing their problems and helping them see success, you will stand out from the other candidates.

Wrapping Up the First Interview

It's always appropriate to thank the interviewer for their time and for giving you the opportunity to meet. Every person who teaches interview preparation will tell you that you also need to confirm your interest in the role. This is good advice because the people who interview you will usually assume nothing about your level of interest. You have to tell them.

Describing your interest is good, but *telling them why you are interested is even better*. The interest you describe should include how the different aspects of the job connect to your value systems. In this initial interview phase, this does not require significant depth, but it does require a genuine connection to prove your value to them, which is the point.

Hopefully, you have noticed all the subtle ways that you can positively exceed expectations in situations where they are not expecting you to. When you excel in places where they expect to get a standard answer, the impact is magnified when they decide to move you forward.

Now that you've read the previous chapters, you should be starting to recognize the art and science of how we find what we want and influence the people who can give it to us.

The Next Level in the Interview Process

Congratulations! You've succeeded in the first round and are headed to the next level of interviewing. We refer to this as the next level instead of the next step, because the number of interviews at each level can vary greatly by employer. We have one client who makes hiring decisions from a single 90-minute interview, and we have others who take three to five interviews to complete. However, the initial phase of interviewing is filtering, regardless of how many interviews there are.

The next level of interviewing is deeper and more personal. At this level, your interviewer wants to know you better. You will almost always see your interviewer face-to-face, whether virtual or in person. This also gives you the opportunity to get to know them better. At this level, interviews are usually longer than 30 minutes. They may not be hours, but they are often considerably longer than the initial interview, and they may include multiple people at one time, a series of meetings, or some combination of both.

Going back to the basics, first impressions matter here, too, especially if this is the first time the employer will see you. How you look, including your face (smile!), clothing, hair, posture, body structure, and more, will all be observed, and some kind of judgment will be made. Our general rule: upon seeing you the first time, the interviewer should just think *sharp appearance* and move on. If the interviewer is still thinking about the way you look 10 seconds after meeting you, there is a problem. This is why we always recommend a neat and muted appearance.

What does that mean? Neat is obviously groomed, with cleaned and pressed clothing appropriate for the work environment. Muted is covering or removing distractions such as tattoos, gauges, piercings, jewelry, and the like. It's also dressing in clothing with colors and patterns that are very subtle. **The goal is that you want your character and qualifications to speak most loudly and not the way you look.** I've never seen tattoos, piercings, gauges, or loud clothing win a job, but I've seen it cause people to lose more opportunities than I care to say.

The military teaches us how to have a proper appearance. They give us regulations on how to press a shirt, how to wear a cap, and what a haircut is supposed to look like. In other words, we're taught how to look sharp and uniform. It's one of the things companies like about hiring veterans and what they hope to see when they get a resume that shows military experience. It is the stereotype, but stereotypes drive expectations, and expectations color first impressions. There is an expectation in this person's mind, a stereotyping of what it is that a veteran looks like. If you meet that stereotype, then you're meeting expectations, and that's a positive. If you are not meeting that, then those things can be negative.

Now, there are also some negative stereotypes related to the military model. Among these is rigidity. This can manifest in a number of ways. Sometimes, it comes from being a little too military—too high and tight, a little too straight-backed, with a blocked chin, and too cold in the process. Too many "yes, sirs, no, sirs," "yes ma'am, no ma'ams." It makes you look like you're not comfortable with people, don't know how to talk with them, and don't know how to work with them.

Sometimes, this rigidness carries over to your work processes—doing everything "by the book" or only one way. These are things that make people in the civilian space uncomfortable, and they feel like these military attributes can divide their teams instead of building them around each other. So, we want to make sure that we meet all the positive expectations that people have about veterans and put them at ease by not exhibiting any of the negative stereotypes associated with military service. Even Montel Williams, a Navy and Marine Corps veteran, intimated to me during my recent appearance on his hit TV show, Military Makeover Operation Career, that he had to be pulled aside early in his television career by a producer to tell him not to bark orders to his staff and say "please" and "thank you."

When I first got into military recruiting in the 1990s, we were very, very careful not to use the term "veteran" because it brought to mind a lot of disabled and broken Vietnam-era veterans to a big portion of America. At that point, that portion of American veterans had been very vocal. In recent years, that stereotype has changed, as most Vietnam veterans have moved into retirement age and are no longer in public view. The result is that "veteran" no longer has that same connotation with much of America.. It's important to understand how monikers, cultures, and brands are created. Like it or not, being a veteran is a brand that has an image, and your first impressions will be judged against people's perception of the brand.

Beyond appearance, the sound of your voice, the feel of your handshake, and any sounds or smells that come from you are observable "appearance" items. Handshakes are usually the most memorable of these. We're looking for what I like to call the "Goldilocks" handshake—not too hard, not too soft, but just right. The good handshake meets the other person at a similar grip strength and rhythm. This is not the time to establish dominance or submissiveness. Equal stature is the goal, with some firmness for projecting confidence and palm-to-palm hand contact. The "limp dishrag" is never appropriate, and neither is the finger-grab.

Other first impression items include open body posture—

meaning hands away from the front of your body and palms turned slightly forward. Avoid hands on hips, arms crossed in front of your chest, or parade rest. Those are all either considered dominant or defensive postures. Instead, you want the posture where you have to shake somebody's hand, give somebody a hug, or something like that —a welcoming posture.

If you are sitting, move your body forward to sit on the front of the chair, slightly leaning forward. This position implies you are paying attention and ready to go. Leaning back in a chair makes you appear far too casual and projects a lack of seriousness.

Finally, there is eye contact. Maintaining eye contact with your interviewer conveys that you are comfortable with the person and honest. Looking down, which is common in Eastern cultures, conveys subservience. Though it may be intended to convey respect, it communicates a lack of confidence in American culture. Looking left or right conveys evasiveness, instability, and or inability to focus, so avoid that as well.

Good posture and body language will help you convey the right message to your interviewer.

Balance of Power

Before we go too far into the interview process, one concept you need to understand is what we call the "balance of power." The balance of power is the measure of who is in the stronger position at each phase of the interview process. In the beginning, this will always be the employer because they have the money, benefits, and responsibilities for the risk of taking on a new employee to support. Knowing where you are in the balance of power equation tells you how to answer questions during the interview and what kinds of questions to ask.

In the early phases of your interview process, you have little relationship with the employer. Most of your influence is based on the potential they can see from your application and resume. Though of small strength, this is what gets you to the first phase. From there, the employer's interest in getting to know you more either becomes stronger or weaker based on their interactions with you and your personal "brand."

This is why your interest in solving their problems has to be emphasized on the front end. If your qualifications are suspect and their perception of your interest in helping them is weak, then their level of interest in you will be low. If they withdraw, they won't feel like they are missing anything important, but you will have lost a critical opportunity to meet your goals on a tight transition timeline. In other words, you have to put yourself in the best position to win every time you can, because you have very limited time.

On the other hand, your interest in solving their problems, combined with your helping them see the ease at which you can transfer your skills quickly to become proficient, will continue to build the employer's interest in retaining your services, and the balance of power will move in your direction.

If you are effective in this, you will notice a change in the employer's demeanor. Instead of holding you at a distance, the employer will start talking with you like you're already on the team, saying things like, "Here is where you will work, and this is your parking spot." These are great signs that your efforts to build their confidence in you are effective.

Once they reach the point of deciding to make an offer to you, the balance of power will have crossed over to your side. The change means that they are now vested in your saying "yes" to them. Should you choose to decline (presumably because you have a better alternative), then the decision to stop the process hurts them a little more than you.

At this point, you can ask all the things that are important to you and your family in this process, and they are likely to be happy to talk about them. This is the point where they want to satisfy your needs because they have decided they want you. They may even modify some parts of their policies to be more accommodating to you. You should feel honored when this happens. Just don't overplay your hand. The balance of power never moves too far on the employee side because the company still has the money, benefits, etc., that the employee needs, but a good employer-employee relationship maintains a healthy balance built on mutual respect.

This also changes how you answer their questions. As the balance moves closer to your side, the employer wants to feel like they are getting the "real you." If your answers feel canned or guarded at this phase, it diminishes trust. There is a general sense that people should open up more as they get to know people they will work with every day, but as just discussed, don't overplay this, either. Become more open, but remain your professional best in everything you do. Never take the bait on foul language. The rules of politeness always apply. The balance of power is dynamic and can change quickly if the chemistry between people changes, but when done right, building a relationship with your new employer should feel very natural.

Back to the Next Level Interview

To understand what gets employers excited, you need to understand the scale they use to evaluate you. To put it simply, the process moves as follows:

- **Can Do the Job**
- **Will Do the Job**
- **Won't Create Problems**
- **Loves the Job and Gets Energy from It**

Establishing that a candidate "can" do the job is simply validating the basic qualifications. Determining if the candidate "will" do the job means that the candidate demonstrates character qualities consistent with being a good employee in this kind of role and has enough interest to accept. Then, the employer evaluates warning flags about the candidate's behaviors and attitudes to determine if they are likely to create new problems as a result of hiring and placing the person in the role.

Finally, the best employees are those who love their job and get energy from doing their job. This means that their value systems are aligned with their role, team, and culture. People who get energy from their jobs require little supervision because the positive feedback they get from performing their role makes producing good work an enjoyable experience. As a result, these people are highly correlated with top 15% performance and long tenures with their organizations. No company can have too many employees at this level, and if a hiring manager believes this is you, your likelihood of being hired is very high.

Notice that none of these explanations included who was the most qualified candidate. When all other factors are equal, and positions are scarce, qualifications and upside potential play a bigger role in making hiring decisions. But at the time of this book's publishing in 2025, there are far too many open jobs in America and not even close to enough people working at the top of the continuum. This is how you win the job and swing the balance of power far enough in your direction to get the best offer, and whatever the organization invests into getting you up to speed, they can now see that this investment in you is likely to be returned to them in the form of strong performance and long-term retention.

Beyond that, people who are really good at what they do are usually promotable in some way. Sometimes, "promotable" can be in the traditional sense, as in the military, leading to supervisory and management roles. Frequently, "promotable" in the civilian space means moving to a higher level of expertise and becoming somebody who uses leadership skills to impact a broader group of people but inside the same functional area.

So, as you are moving through the next level of the interview process, you should now be more able to understand what you are trying to accomplish. Whether many people interview you or only one, the better you can move people through the continuum in their evaluation of you, the more likely you are to succeed in gaining employment with that organization.

If more than one person interviews you, here are some roles other than the hiring manager that you may have to influence and what is frequently most important to each of them:

Human Resources. HR handles most of the complaints within an employee organization. Most complaints they receive are about the way that employees interact with each other up, down, and across the organization. Therefore, when meeting with HR, giving them confidence that you can work with all kinds of people for the good of the organization, show respect to everyone, and work hard to resolve conflict will help you get the up-check from this important meeting.

Peers. Peers are used primarily for technical evaluations and to determine fit with the team. Technical evaluation criteria are usually pretty clear, but fit with the team is important and more nuanced. At the core of all of this, most people want to like their coworkers and trust their work. The core of liking in the workplace begins with willingness and initiative to help teammates, and from there, it grows to confidence in your competence. Peers are usually most critical of those who can't learn and apply knowledge and who repeatedly ask questions about things they should already know but don't trust their own judgment to execute. Address these issues, and your peer interviews are likely to go well.

Executives. For interview purposes, the executive is the manager above the hiring manager. Executives like to see their managers making good decisions in bringing great people to the organization. Executives frequently want to make sure that candidates know the job and that it fits with them. They validate that the candidate fits the company's culture and provide another checkpoint for any warning flags. By being interested in being a part of the team and helping the executive meet their goals, you are likely to be successful in winning this most important person to your cause.

Changes in Modern Management Approaches

When I started in recruiting in 1999, Jack Welch was CEO of General Electric and was widely regarded as the most successful CEO of the last half of the 20th century. He took a nearly 200-year-old company full of bureaucracy and dead weight when he started and turned it into a highly competitive and profitable manufacturer that became the envy of the world. Welch believed in the military model of pushing authority and responsibility as low as possible in the organization. By the time Welch retired in 2001, everyone in the world was trying to follow GE's lead.

That didn't last long, and the reasons are simple. Human beings don't want accountability for their decisions. This has been true since the Garden of Eden. As time has gone by, placing decision-making authority as low in the organization as possible and holding those decision-makers to account has quickly given way to decisions having to be made at higher levels and only with the concurrence of multiple people. Though some parts of this change have been proven positive (particularly when it comes to analyzing risks and alternatives), the negative side has been that accountability for bad decisions is hard to find. Observing recent events, even the armed forces haven't escaped this trend.

Management still has the right to override the inputs of its people, but it requires the manager to accept the risk of accountability. That is why you have to build consensus for your being hired with everyone you meet, and the larger the organization, the more consensus you will have to build.

Tripwires in Interviews

"We"

Part of military leadership training includes passing credit for success to your team and taking personal responsibility for failure. These are the things we do in public or in front of our troops, and it is absolutely appropriate to do so in those situations. However, when it comes to your evaluations, resume, and job search, your accountability for failures gives you the license to take credit for your successes in these situations. If you would have been held to account for failure, you need to take credit for the success when you are interviewing and writing your resume.

Additionally, your interviewer is not going to hire your team. Your interviewer is evaluating you alone. Passing credit to your team when answering questions is fine, but your job in taking credit for the successes of your team is to describe what you personally did as part of making the success happen.

With military technicians, however, we sometimes see the opposite. In these cases, a junior person takes credit for a success that they did not create on their own. When I ask about difficult equipment troubleshooting situations, we frequently hear from junior techs a story about a group troubleshooting effort in which the junior tech participated. The junior tech claims victory for solving the problem as part of the story because of its difficulty, but the story reveals that he was only an observer.

The tech's participation in this story could still be good in the context of the success, but the story has to be changed to highlight what the junior tech actually did to contribute to troubleshooting the system and restoring it. Again, the interview will not be hiring the whole team but the individual tech appearing before them.

Brevity

There are two times in interviews when a person can literally talk himself out of a job. The first is on the "Tell me about yourself" question. Several times during a debriefing with my candidate, I asked how the interview went and got the following answer: "It went great! They only asked me one question." In these cases, my follow-up questions always reveal the same thing: the candidate talked for the entire interview and never gave the interviewer a chance to ask another question. The debrief with the employer usually proves equally predictable: "No."

Not Enough Detail

It may seem funny to talk about not saying enough as the second thing that can cause you to talk your way out of a job right after the section on brevity, but like most things, the goal is balance. The things you say need to have enough detail to be complete but not so much detail as to run unnaturally long. The key to adding detail is relevance. If the details are relevant to the question or would be part of what a person in the role would do, then they are worth discussing.

If you are interviewing for a technician or engineering position, the most common technical question that veterans fail is: "Tell me what you did in the service." Without training and awareness, the veteran will cover the full spectrum of what we do on active duty with very little depth. That may be what the folks back home want to hear, but your interviewer ONLY wants to hear about the things you did that are relevant to the role you're seeking. This is why it is so important to take stock of the functional duties you performed. You will need to talk about them in detail and with specifics.

You may say, "But Mike, my work is classified. I can't talk about it." Let me be clear. I would never suggest that you compromise your security clearance by disclosing protected information on a resume or during an interview. This is why we focus on functional duties.

Antonio Mendez wrote a book titled, *The Master of Disguise: My Secret Life in the CIA*, published in 1999. Mendez was a CIA agent who used disguises to infiltrate some of the most hostile places in the world and hide intelligence gathering systems, frequently in plain sight. As you can imagine, every operation was highly classified, but he was able to write an entire book on the work he did by talking about the functional parts. You see, the basics of whatever you do are not classified. It is how those functions are organized and employed that makes them classified. If you're an electrical engineer, electrons are not classified, but if you convert those circuits into cutting-edge listening devices used in spycraft, the application of that technology will be classified. Your operation may have been classified, but your approach to planning and executing it is likely not and can be explained that way.

The bottom line in all of this is that you have to provide details on what you have done if you want that person to pay you or endorse paying you, because they won't accept "Trust me; I have a security clearance" as a good answer.

Adding Parts to the Question That Were Not Asked

Most interviewers are looking for simple and straightforward answers to their questions. If you are an expert in your field, you likely understand levels of detail and nuance that the average person who interacts with your topic does not. Where I have seen this cause problems in interviews is when the veteran inserts these into questions without clearing it with the interviewer.

If you are an expert in your field and think that your answer may require depth, ask the interviewer about the depth they want. Unless they are asking for a story, they usually want an answer that can be covered in less than a minute. This is not the time for you to insert your brilliance unless you know you are talking with another expert on the subject and it's relevant to the role. Otherwise, the interviewer will lose interest quickly and stop listening.

Here's an example to illustrate this. A number of years ago, I was working with a young West Point graduate who graduated at the top of his class and was Chairman of the Honor Committee. This senior leadership role in the Corps of Cadets is the only role that is selected by the senior class itself. In other words, the person selected for this role has leadership and personal qualities that are well-known throughout a class of about a thousand of the nation's finest leaders. He was impressive and exhibited all the best qualities of a young Army officer.

This young man found himself in the unfortunate situation of having graduated from West Point but being discharged without a commission because the Army needed to reduce the officer head-count. When that happens, they frequently make their initial cuts by revoking waivers for any kind of medical condition, and that is what happened to this young man.

In helping this man find a civilian job, I introduced him to a food manufacturing company that routinely hired new college graduates for production supervisor roles as part of a management training program, and they had hired more than 20 veterans from me—mostly mid-grade enlisted candidates who had already earned a bachelor's degree before their discharge. My candidate was motivated by the role and the company. I was well-acquainted with the company's interview questions, and he prepared and did everything I taught him to be successful.

In his first interview with an internal recruiter who was only a year or two older than he, my candidate was asked a scenario question of how he would handle an underperforming hourly associate. This young graduate was an expert on leadership and gave her a brilliant answer full of great insights and scenarios. In my debriefing, when I reviewed the question with him, I was really impressed by his answer and thought he crushed the interview and was sure that this would end up a slam dunk.

When I debriefed the interviewer, I was absolutely stunned by what I heard. Our stellar West Point graduate, top 5% of his graduating class from an institution that is world-renowned for developing the best junior military officer leaders in the world and selected by more than a thousand of his peers for the only peer-selected leadership role in his class, was declined a second interview for a job that is routinely filled by middle-tier new grads from middle-tier universities across the country. The feedback: "Insufficient leadership skills." Stunned was not a strong enough description for my reaction to the news. I was flabbergasted.

My candidate was an expert on leadership. I am an expert on

leadership, but the interviewer was not. She only needed to hear three points to move him to a site-visit interview, and she didn't hear them. He talked so far above her head that she missed the simple three items that needed to be covered from the question on her interview feedback form for him to move forward. For all our expertise, she was still the one who controlled going to the next step, and she said "no."

All she needed from my candidate were the qualities of a leader that you learned in basic leadership training. You take time to know your people. You recognize a performance issue. You address the performance issue with the person. You create a plan to get the person back on track. You follow up on the plan, and the person's performance is restored. She couldn't recognize three of these points in his answer, and that's why he scored as having "insufficient leadership skills."

Stories That Miss the Point

This tripwire frequently occurs in situational interview questions. A situational interview question requires you to tell a story, making it easy to get off track and forget what you were supposed to discuss in telling a story about something you did in the past. That's the tripwire. Before answering a situational question, select your story by remembering where it has to end—happily. We recommend telling your story in four phases and using the mnemonic device "SPAR" to help. If you've been through TAP class, you have likely heard this as "STAR," but we have found that a bit confusing, and we like the boxing imagery that goes with ours because your story has to land with impact. The four main components are:

S—**Situation:** The situation should tell us when, where, what, and who is involved in the story. It also needs to lay out the challenge and help the listener understand the potential for failure and its impact on the organization or its mission. This is the most significant part of the story because it pulls your listener in. It should be 40–45%

of the entire time of your story (45–55 seconds for a two-minute story, 70–80 seconds for a three-minute story).

P—Plan: Before addressing any situation, everyone has some kind of plan. That plan may change, but at least be sure to include three initial steps in laying out how you intended to overcome the situation you described.

A—what was **Actually** done: Successfully overcoming any challenge requires action. Tell the interviewer what you *actually* did. At the same time, every military planner knows the maxim: "Every plan looks great until it comes in contact with the enemy." Frequently, circumstances change or assumptions prove incorrect, and your plan has to be changed on the fly. This is where you would talk about that. What was actually done is the second most important part of your story, because this is where the details are to qualify you, but it's also where the drama is that makes your story interesting. (About 40% of your time will be spent here.)

R—Results: This is where you have to land your story for impact and show how you created success from the situation. Keep this front of mind the entire time you are telling the story. If the value of the results is not obvious, then explain to the interviewer why success is important to the organization. You may have to revisit the implications of what would have happened if you failed to help your listener understand the significance. With active military, the different inspections and unit certifications frequently fall into this category. For example, an Army leader may say that the result of their actions was a successful NTC (National Training Center). That won't mean much to a civilian, but explaining that you have to be certified to deploy at NTC to perform your mission overseas implies that failure can lead to the firing of the commander, and all civilians definitely understand why that would be important.

Remember, your story should last two to three minutes, based on its significance and complexity.

Understanding the Power of Story to Influence

The power of storytelling is so significant in interviews and human existence that I'm taking this section to help you understand why that is.

Every story has to start with a setting. The setting puts the story in context, which is a fancy way of saying that it helps us understand the starting point and what is going on. Now, if you are a gamer or someone who likes adventure books, this is the part of the story where the hero's challenge is revealed. In your interview story, you tell how the hero (you) was confronted with a challenge where their decisions and actions led to success and victory. When you tell your story, you have to make sure your interviewer understands the extent of your challenge, and the only way to effectively engage your audience and paint this picture is to give them enough detail.

As a matter of fact, the reason why this is the most significant part of your story (and answer to the situational question) is understanding how important storytelling is to human existence. Without going into a great deal of detail, there is a book written by Michael Bosworth and Ben Zoldan titled *What Great Salespeople Do*. The authors describe their book as the science of selling through emotional connection and the power of story. What's significant about this book is not that it's one of the many thousands of sales books that are out on the market today. The significance is that it's a follow-up to a study that was done in the 1970s and 1980s related to a book called *SPIN Selling* and the companies that were spun off from that study that consulted sales teams across America and around the world.

The first thing that the authors say in their latest book is that what they were trying to do was completely wrong, showing an honesty that I found remarkable. The research behind *SPIN Selling* showed that 20% of the salespeople in any organization produced 80% of the results. Developing the *SPIN Selling* model was supposed to help the remaining 80% of salespeople contribute more to their company's sales. The idea was to give more tools to salespeople so the 80/20 rule of sales production would no longer apply.

SPIN Selling uses a consultative selling process through the use of different types of questions designed to lead a prospect through a process of guided discovery, culminating in them developing their own logical conclusion to buy. Though *SPIN Selling* was successful in many ways, the authors discovered that it failed to meet its original objective. In fact, it made it worse. Among the companies that implemented the model, the top 20% went from producing 80% of sales to 87% of the sales.

This discovery compelled the authors to completely reevaluate everything they had ever believed about improving sales performance and start over with a whole new study. What the new study revealed is that all the most effective sales people demonstrated the ability to influence prospects by telling them relevant stories. This led the authors to look at research on how stories affect people's brains. What they found was that live brain monitoring showed increases across far more parts of the brain than any other way that people interacted with information and other people. Technology explained what the observations had shown.

The economic value of storytelling proves this point even further. Think about how much of the world's economy is built around storytelling. Certainly, we have Hollywood movies. We have television shows. We have novels. We have magazines. We have talk shows, and we have news programs. Millions of stories are being consumed all day, every day, and it has become abundantly clear that the human mind has been engineered to consume stories. As a matter of fact, we know that this has been true for most of history, as stories were passed down from generation to generation through oral tradition before we could preserve knowledge in writing.

Here's something else that we all know about stories: The best stories make us feel like we're inside of them, a part of them, present as we're watching them unfold. Whether we watch a story on TV, read it in a book, or hear about it on a radio broadcast or a podcast, ultimately, stories trigger the different senses and emotions in our brains. We don't see with our eyes; the eyes are the sensor. It's the gateway to the brain. We actually see in the brain. We don't hear with the mechanism of the ear. The ear is the sensor that captures the vibration signals and translates them to the brain, where they are processed as sound.

It all happens in the brain, and just as neurons transmit signals for smiling and body movements connect to the emotional centers of the brain, research shows that the same responses are triggered with words.

When we hear words that trigger the senses in the brain, they become real and alive. And when you can connect stories to people's senses and their emotions, you are making them come alive in their minds. This is why you can watch the same movie 20 times and still feel your heart rate rise at the moment of confrontation with your hero, even though you already know how it's going to end. Your body is reacting to those stories and to what's happening as if it were being experienced for the first time. The same thing will happen with the stories you tell. This is why laying out the situation at the beginning of your situational question is so important.

You need to help your listener come into your world and see the story through your eyes. Now, here's something else that's interesting. If I opened my story by saying that I found myself at my military unit, in my rack or my bunk, and I was startled and turned on the lights, every person reading this has had a different experience about what that's like. If you're not military and you're reading this, you have a different understanding of what that looks like and what that experience is like.

The important thing in storytelling isn't that your listener "sees" the exact same thing that you do. No, the important thing is that they actually see something in their minds when you're telling your story. For example, if I describe walking down a long passageway, most Navy veterans will see a particular color of blue tile on the floor and white padded walls with pipes overhead—a typical passageway on a

Navy ship. When I tell them that there was a door at the end of that passageway, they're going to see a door that has a unique type of hinge on it with a large lever handle or multiple smaller levers that are used to seal that door from water in the event that the ship is in danger of sinking or spreading fire. If I talk about a passageway in my story to a government employee, I suspect this person is going to see a hallway in a building with a very typical door down at the end of it, because that is what is familiar.

The actual images in the mind of the listener don't matter, because we're succeeding by stimulating the person's brain to engage with us and become a part of our story. So, when we lay this out, our listener has to hear enough details of the who, what, when, and where for their mind to create a picture. In other words, give them a view of what you, the hero in this story, are seeing around them. This will bring your story to life in their mind.

Again, this is going to be 40–45% of your story when answering a situational interview question, and most of these situational stories are going to last between two and three minutes. For example, you may be describing how you had a very long workday, were exhausted, went to bed late, and were woken up in the middle of the night to address an issue. Adding description to what made the day so hard is relevant and puts your state of mind and physical condition into context. Perhaps the temperatures were extreme while you were out on patrol, working on an aircraft on a flight line, or laboring in an engineering plant. Anything that makes our military jobs harder is relevant when it comes to understanding our hero's challenge.

These details help your listener relate to limitations that may have made success harder to achieve, potentially increasing the risk of failure your efforts were engaged to avoid. It makes the story more interesting and causes people to feel the fear and anxiety that our hero was going through. At the end of the situation description, our listener should know clearly what our hero was facing.

From here, every story creates a drama that unfolds as the hero sets out to address the challenge. In interviewing, the first step in building this drama is the "plan." Your plan gives the interviewer a little bit of insight into how you tackle problems.

The drama builds as we move into what "actually" happened. It's the part of the story where engagement is happening, and more vivid details bring your story to life and make the sequence of events more interesting. Details of your actions relevant to the role you're seeking help your story. This is where they are evaluating your capabilities, but you're really reeling in your listener at this point. Your interviewer should be following your story like they are in it with you. Your words help them feel the same things you did, and at a certain point, the tide turns in favor of our hero. When we hit the climax of the story, your interviewer will start to feel relief as they see a successful ending coming. Again, this is about 40% of an interview story.

From there, the drama starts to wane, and the story is concluded. This is where you introduce the results of your conquest and your happy ending.

As a side note, when I'm coaching a military leader on handling the question of how they dealt with an underperforming subordinate, it's far too frequent that we get to the end of the story, and they say, "Yeah, and then we got him booted!" If you've been a military leader and been through this process, you understand why this is such a big deal, but that's not what our civilian counterparts want to hear from this question. Though it may be true that you could be asked a question about when you had to terminate an employee, this is not what they were asking. If you ever need to provide a negative result, you will be asked for it specifically and clearly. Otherwise, assume they want to hear a happy ending.

In other words, the stories have to have a happy ending. That's what employers are looking for in nearly every single case. Now, in the military, we sometimes have results that are positive and significant in ways that aren't obvious to our listeners.

As you tell your story, keep it moving quickly and make sure it has a logical path so it's easy to follow. Also, this is your story. Tell it with enthusiasm. Even if you don't consider yourself to be a dynamic storyteller, by working this through, you will have a story that will come in, once again, in the Goldilocks space: just right.

Finally, remember this: Stories that we connect with and that stir our emotions influence our thinking. Don't believe me? Think about testimonials that people have made to you about a bad experience with a product or service. How did it shape your opinion? The same applies to the positive. Someone tells you a story about a positive experience with a product or service, and if that story connects with you, your opinion of it immediately improves.

That is why so much of the world's economy is tied up in stories, and the more vivid the story is (triggering senses and emotions), the greater the impact it has. The idea is to influence your interviewers with stories that give them a high opinion of you and your abilities so that they want to hire you.

Summary:

Interviews with employers are where the magic happens. Remember the following to help keep you on track.

1. First impressions are vital. Remember, the good military veteran brand is that of a professional. Meet or exceed these expectations, and you are on the right track. Always be your professional best in what you put out to the world, with special attention paid to your social media. Expect an employer to find you online if you have ever had social media.

2. Initial interviews are almost always done by phone or video and are used to filter candidates to make sure only the appropriate candidates pass through. The goal is to get through to the next interview. There are some organizations that will make their decision from a single interview, but in nearly all cases, the initial interview is done on-site. If the job is for a remote position, video may

be used. And, of course, expressing enthusiasm will cover a lot of mistakes.

3. "Tell me about yourself" is not a personal history. Sometimes it's asked in different ways, but ultimately, it is the point early in the interview where the employer asks you to talk about your background. The interviewer is only listening for details about your training and experience that relate to the job. Make it easy for her by providing relevant details of your training and education combined with specifics of how you applied those in your work life (functional duties). If you have extended your training and experience to help you solve problems or deliver results where you've had no training or experience, that is even better. Just remember, be brief. Giving this answer should take no more than a minute or slightly more in most cases.

4. Ask your first question immediately after covering "tell me about yourself," which focuses on the expectations a person in the role will have to meet to be successful. To really make an impact and position yourself as an expert, use something from your past experience as the lead-in. For example, in a case of interviewing for a team leader role, you may say, "It's been my experience that leaders who relate well to their teams frequently get the best results from their people, what kind of leader do you feel will get the kind of results you want from this team?"

5. Different kinds of questions will require different answers and approaches. Questions that ask you to "tell me a time when" are looking for a short story, bring it to life with SPAR. In all cases, remember your answers should have happy endings in nearly every case.

6. Understand the Balance of Power in an interview. Always remember that the only reason an employer gives you time is to solve their problem. In the beginning, the

interviewers will only be listening to what's important to them. As they vest in you, they will be increasingly more interested in what is important to you. Setting the hook in the beginning by focusing your interests on solving their problems will start tilting the balance towards you, and that is how you get what you want.

7. Be sure you answer the question that was asked. Don't add anything that wasn't asked unless you know you're talking to an expert in that field who would be interested in the nuance. Be brief but complete. Being brief means giving all of the important details in a short period of time. The key is to remember that details still matter even when you are keeping your answer short. Speaking in generalities conveys that you are uncomfortable discussing the topic, and in an interview, it means that you don't know it.

8. Stories help you connect with and influence people. Using SPAR to guide your story will help you stay on track. Be descriptive in setting up the situation to help your listener engage. Help them visualize the hero's challenge and make them feel a part of the quest to reach victory.

9. Winning in interviews means being viewed as the person most likely to succeed in taking away the pain they know they have without creating new pain for them.

Chapter 7
Technical Interview Challenges and Closing

During technical interviews, you're sometimes going to encounter things you don't know. In many cases with military veterans leaving active duty, these things won't be used to disqualify you, but instead, they are asked to help the interviewer understand the limits of your knowledge or what level of training or exposure may be necessary to get you acclimated to their systems and products.

The real question is: What should you do whenever you run into a technical question that stumps you during an interview? It's been my experience that treating the question as a troubleshooting problem gets the best results. As previously mentioned, an interview isn't usually like going before a military board. It's not like a class and taking a test, where you can't ask questions of the person asking the questions. This is a consultation at every phase of the process. So, unless you're told that you can't ask questions, assume that you can.

This means you should treat the technical portion of an interview like a troubleshooting problem. The key to troubleshooting is starting with the things you know and working your way through the process to try to find the answer you don't know. This is a particularly important process to use during a technical interview, whether we're talking about engineering, information technology, or anything similar.

Most technical interviewers and professionals very much appreciate people who exhibit curiosity, and curiosity in the technical space manifests itself through questions, interests, and what many people describe as a desire to solve puzzles. So, feel free to troubleshoot what you're encountering out loud. Your interviewer will most likely appreciate hearing the thought processes that go into this, and you may just end up finding the right answer.

As your interview moves along, you're going to find that it has ups and downs. However, as long as you're keeping it moving along and don't have any of those moments that would shunt you to a no, the interview should feel very comfortable to you.

Closing Time

We talked in the beginning about how first impressions are lasting impressions, so it's important to get your interview off to a good start. That is why we focused on making sure your appearance, your presentation, and all the things that people make snap judgments about are good and well put together. Additionally, for military members going into the civilian world for the first time, a good start to an interview is vitally important, because most find it difficult to get an interview on track if it doesn't start on a good note.

The second most significant thing that people remember about meeting a new person for the first time is how the encounter ended. Like most impressions, it is the feelings about a person that are remembered. So, the end of the meeting can affect the outcome in the same way, either negatively or positively. When it comes to interviewing, volumes have been written on making a good first impression, but not nearly enough attention is placed on ending impressions. At best, people who teach interview preparation suggest asking for the job or at least the next step in the process to help create a positive inflection in the interviewer's opinion. The problem is that it is so common that interviewers are expecting that, and therefore, its impact is attenuated..

So, we offer our military candidates an important way to make as significant an impact on the interviewer at the end of the process as they did in the beginning. As we mentioned, there is an expectation that you need to tell the interviewer that you are interested in the next part of the process. If you do that, you've at least checked the box. But if you want to really make an impression on your interviewer, you need to tell them why, and there is both an art and a science to how you should roll that out.

At the time of this book's publishing in 2025, we have been living in a world where jobs are plentiful, and good people are few. After more than 10 years of strong increases in the jobs market, employers are willing to keep jobs open much longer to find the right person, and they are more willing than ever to train people wherever they can afford it. But, they also understand that if this job isn't a fit, you're going to continue to look and ultimately take another one somewhere else and leave them with nothing to show for their investment of time, money, and resources in developing you.

So, as we start to plan for a strong finish, let's review how we got here. The person is sitting in front of you because they have a "pain." They have a problem that requires a person to solve because they currently do not have someone on staff to do the things they need. In other words, the job, its requirements, and the environment where the work will be done are critical to making this person's life better.

In the search and placement side of our business, we fill a lot of field service engineering roles for our clients with veterans leaving active duty. Frequently, in a field service position, the role requires a person to be living in a particular location or region based out of their home. The job requires travel to customer locations by car or plane, and there is usually some type of training (investment of time, money, and resources by the company) that goes into this in the first few months of employment.

Though we effectively match our job-seekers with clients, we help them understand that they need to be sure to tell the employer what they like about all the different aspects of the job and to focus on the most critical elements first. It's critical that they like the location and have a reason to be there. It's important that they like the idea of travel, going out to different places and working in different places. It's critical that staying overnight in another town is considered a positive aspect of the role. The interviewer also wants to know if training is a positive because it will be a significant investment. It's also important that the person likes the work: installations, troubleshooting, and repair.

When these items are the first things you say to the employer who has a field service role with an overnight travel requirement, you build their confidence that your interests fit for all the right reasons because they address their source of discomfort or challenge they are trying to overcome. They want to know if you are truly interested in the job, including the parts that may be challenging or arduous. This is so important that it should be about 60% of what you are going to say concerning your interest in the job.

The second thing that builds their confidence is you validating the people you have met through the interview process. Among the key things that employers look for is how well you get along with people. In fact, you'll find that most of the time, people are terminated or prove to be bad hires for reasons that have little to do with their qualifications. It usually has to do with their ability to work with people and the character qualities that they demonstrate in relation to the requirements of their job.

So, after you've explained all the reasons why you like the job, you want to talk about the people you have met along the way and what you value about them. Perhaps it is that manager whom you feel you have a good rapport with. Perhaps you can see that this is someone you can learn from. You can mention that you enjoyed the conversation or that you have things in common, any number of those things. But the important thing is to validate the people. Tell them that you have appreciated them, enjoyed them, and feel as though you fit in with them. This should add up to about 30% of what you say you like.

The last part you must like is the organization itself. This means talking about the company, agency, or organization you would be working for. These comments can include the product, success, executive leadership, culture, reputation, benefits, etc. This should account for roughly 10% of what you say, making it fairly brief. It's good to have pride in working for a company, but it's the least important thing overall.

What is interesting about this is that if we leave people to their own ideas on approaching this, they will almost always talk more about the organization and the company, maybe a little bit about the people, and almost never get around to the job. How do we know this? Because this is where the employer's people usually start when we ask them about reasons to take the job, as these are the easiest things for them to talk about. However, it isn't what is most important to them in hiring someone, and it's also not the most important reason for the job-seeker to accept the position. Ultimately, both sides are interested in the fit with the job first, and then the people, and this technique leverages that.

Ultimately, the work that you do every day, where you do most of it, and the people you do it with are where most of your work-related happiness and satisfaction will be derived. Therefore, we want to spend 60% of the time talking about the job and things you like, 30% about the people, and then just a little bit about the organization.

Here's how this may sound as you make your closing the interview: "Tony, I really appreciate your taking the time to meet with me today, and I have to tell you that I'm really excited about this opportunity. Let me tell you why. First of all, this position is located in Pennsylvania and close to my family, and I love the fact that I will be traveling out to customer sites and getting to see the country as part of my work. I've always loved maintenance, and the idea of being trained on maintaining new systems is exciting to me, but I'm also interested in learning about how to install systems in customer facilities. So, the job looks great to me! Additionally, I've really enjoyed getting to know you in this process. I can see that I can learn a lot from you, and I feel we will get along great. Pepper in HR was terrific. I enjoyed meeting her, and if I get the chance to work with Nick, I think he and I could really work well together. Finally, Stark Industries looks like a terrific company overall. I love the fact that we make innovative products, and I love the company's commitment to its people in more than just words but also in the great benefits package I've seen."

And now that you have them hanging on your every word…

The Magic Move

The "magic move," as we call it, is the statement that transitions you from your interest confirmation to your closing question.

Continuing with our field service example, the magic move and closing question would sound like this: "So, given all of these good reasons, I feel like this position is a great fit for my skills and interests. *Do you have any hesitations about my ability to become the next great field service engineer on your team?* As you can see, this job looks like a great fit for me, **and I'd appreciate it if you gave me a chance to address any lingering concerns you may have before I go.**"

Asking an interviewer this question says so many positive things about you. It would be hard to list them all here, but among the most significant is that it demonstrates your interest in the position and that you care how you are seen by somebody else. How many times have you heard somebody say, "I don't care what this person thinks about me." By asking this question, you indicate that you actually do care about what people think of you in the professional world. That is very, very important.

The second part of this is that you have demonstrated that you are open to feedback. You are asking them if there is an objection to your being hired, which means that if you were to get some type of feedback that indicates a concern, you'd be willing to take that. That is a demonstration of what it would be like to work with you in real life, and it's a very important attribute. Finally, you demonstrate your eagerness for the position by your willingness to put yourself in a vulnerable position of potentially getting negative feedback.

Now, it's been our experience that there are only three ways the interviewer can answer this question. The first of these we call "the confession." The confession is what we want. It is the confirmation of your fit, and it usually sounds like this: "No, I don't have any hesitations at all. I think you would be a great fit for the position." Getting your interviewer to make the confession is important because what we confess with our mouths, we will defend to the end of our days. To do otherwise would be to admit to being a liar. Now, this isn't to say that you are getting the job, but what it does say is that the interviewer was satisfied with what they got from you and believes that you would be a good fit.

This is what you can control in the interview process: getting yourself to the position where you are a finalist for the role. What you can't control is the competition, but the confession from the interviewer that you are a fit is something that will carry forward from there, and you know that your application is still alive with them, which is very important.

The second response you could get is a lingering concern. This is actually a good thing because you will have been given a second chance to recover from a mistake you made. You need to understand that any objection to your employment that has not been addressed will lead to your being declined for the position. A company will rarely hire someone whose interview leaves lingering concerns because hiring someone obligates the employer to significant amounts of money and long-term liabilities. If the employer is not confident in that investment, the inclination is always going to be to just skip over the candidate and go to someone else.

The other bit of good news about the concern is that if you have reached this phase of the process and still have a lingering concern, it is usually not something terribly difficult to resolve, and the interviewer is likely willing to help you overcome it. All of this is good. In most cases, this happens when you didn't fully answer an important question that has been asked.

We have found with military transition candidates that the question that did not get answered properly is usually a situational question, one in which you were asked to tell a story about something from your background, which is why I emphasized STAR and storytelling in the previous chapter. When a person misses on these kinds of questions, it has been our experience that the person delivered a hypothetical scenario answer instead of a story or that the person forgot where the story was supposed to end to properly answer the question and missed the landing spot.

Handling this concern is simple if you implement the following steps:

1. Repeat the question back to the interviewer in your own words so you are clear on what they want you to address. (Again, this is usually going to be a situational or scenario-based question.) Finish your statement by telling the interviewer how you think the question should be answered and where it should land. For example: "So, you would like me to tell you again about a time when I had to lead a team to a challenging goal, and I should describe the goal and what made the goal challenging and then talk about what I did myself and in leading my people to reach the goal. Is that correct?" When you do this, you will usually see/hear a positive reaction from the interviewer, who will likely respond, "Yes! Tell me about that!" The process of clarifying the objective and speaking it aloud has the collateral benefit of clarifying what you need to do and etching it into your own mind.

2. Answer the question the way you planned.

3. Confirm with the interviewer that your answer satisfies

their concern and that they are now comfortable recommending you as a good fit for the position.

Address the concern in this way, and you should now get "the confession," which is always the goal.

The last response you can get to the closing questions is what we call "the rope-a-dope." A rope-a-dope answer sounds like this: "Well, we're interviewing a number of candidates, and we will get back to you in the next couple of weeks." But does this actually answer your question? No, it doesn't.

Don't immediately assume that the interviewer heard your question completely and answered you intently. Our experience has shown that when you get this response to this question, the interviewer simply wasn't listening to you. They were expecting that you were simply asking for the job or asking for the next step in the process, which is what most people are taught to ask, and they gave you a reflex response.

The best response to the "rope-a-dope" is to acknowledge that you didn't ask the question they answered. It might sound like this: "No, no, no, no. I didn't ask whether I had the job or if you were going to take me to the next step. My question was simply, based on our conversation today, do you have any lingering reservations about my ability to be a great fit for the job?"

The important thing to understand about the rope-a-dope comes from something we covered earlier. Employers know at the core level that for them to fill their position, the candidate they want has to say "yes." Therefore, employers who interview job candidates develop the habit of making sure that they answer the questions of the people they're talking to. In this case, the interviewer you're speaking with missed your question.

The reaction you typically get from the interviewer, in this case, is the effective admission that they weren't listening closely enough to you. We all know that it's offensive when you are in a conversation, and the other person isn't listening to you, and that pressure is at work here. Your interviewer will feel as though they may have offended you by not listening to you, and their first reaction is that they will feel like they owe you something for this.

Usually, this is where an apology is made, and right behind it is the confession.

Sometimes, especially if you are talking to someone in human resources or talent acquisition, you may still get more of a non-answer, and I don't recommend pushing it further. Their evasiveness may simply be policy, but in many cases, if you address the rope-a-dope, you'll ultimately end up with the confession of a good fit.

In summary, remember that a strong close that provides detail on your interest in the requirements of the job, validates relationships with people, and confirms interest in the company will get your interviewer's attention. Adding the closing question will likely exceed the interviewer's expectations of you and make a great final impression while giving you the benefit of really knowing how you did in the interview.

Final Note: If you can't bring yourself to ask the closing questions as I've outlined them and find yourself thinking of asking something like, "Do you have any other questions for me?" Don't bother to ask. A question such as this lessens the impact of your well-considered statements about your interest in the position, and that works against making a strong final impression.

Chapter 8
Narrowing the Field

Interviewing is a two-part process. The interviewers have a responsibility to do due diligence—in other words, to do an exploration or analysis of you as a candidate—but at the same time, you are doing an analysis and your due diligence of the opportunity in front of you. As you go through this, you have to be able to decide what it is that you want to move forward with and what you don't.

One of the interesting things about the military transition process that transition offices have taught for a long time is that you should keep your options open for as long as you can. Once you've done your interviews, you go from keeping all of your options open, to moving to close them. After all, you can only accept one job. That can be very challenging for people, especially if they have been thinking about their transition for a while with the intention of maximizing opportunity. As a matter of fact, an element of fear of loss is involved with closing out options and has to be recognized.

There is a fine balance between keeping your options open to create a market for your services and starting to close them out to narrow in on the one you want. The key goes back to how we started this process. It's like climbing a hill. We start with a plan and then do a lot of work to get to the top of the hill. But as we start down the other side of the hill, we bring ourselves back down to a place like where we started at the bottom of the hill, representing our value systems and what we want to do.

When it comes to evaluating your options, look at the opportunity you just interviewed for, scoring it in terms of the value system and your priorities around them that you put together at the beginning of this process. One of the easiest ways of doing this is with a pluses and a minuses chart.

To do this, take a sheet of paper and, on one side, write down all the things that are important to you—location, a particular part of the job description, functional duties, etc. Of course, you should list them in order of importance. Then, on the other side of the paper, write down how the opportunity matches each of those things that are important to you. Finally, give each a score that you can use to easily compare this opportunity to others.

You do not have to be precise. You don't have to lay it out on a scale from one to 10 and try to figure out the difference between whether this is a six or a seven. Really, it comes down to four basic grades:

- **A grades** are for aspects that are "really exciting" and that you find motivating.
- **B grades** are for criteria that are "something that I can do and/or enjoy." It fits what you want, but it's not the most motivating aspect of the opportunity—though it's close.
- **C grades** are for criteria that "satisfies my need in that particular area but doesn't get me excited or move me quickly toward my long-term goal." It checks the box—nothing more, nothing less.
- **D grades** are for criteria that are "significantly lacking." If this particular criteria is central to your value systems, it could be a deal-breaker. At a minimum, it would be

something you want to improve about your job at the soonest opportunity.

If you have prioritized your value system and graded each aspect of the opportunity in relation to it, you'll find yourself being steered toward the things that are most important to you, and you'll find very quickly that opportunities will start to separate themselves.

The interview and transition process may cause you to alter some of your short-term priorities based on time pressure, experience with the process, the number of opportunities you have, personal circumstances, and personal finances. This is OK. Usually, the first time you go through defining, prioritizing, and ranking your value systems, they will not truly be in alignment. Periodically reviewing your value systems as part of the military-to-civilian transition process and then as a life habit will help you get good at this and will lead to a happier and more productive life.

A Word of Caution

Managing emotions can be challenging.

Uncertainty about the transition process can induce fear, which convinces the transitioning service member to open himself to many options he doesn't really want. Getting invitations to interview with many of these can often supplant initial with misplaced arrogance from thinking that a company's interest to interview is a "job offer," which causes the candidate to torpedo other opportunities and shun people willing and often more able to help. The cycle of emotions then turns again as the projection of arrogance causes the veteran to lose opportunities, and as time pressure builds as the end of service date gets closer, arrogance converts to panic. Once panic sets in, the veteran feels compelled to take the first offer they get just to relieve the panic by regaining confidence he will be able to pay his bills. Before long, the veteran realizes he accepted a job he doesn't really like or joined a company whose culture wasn't a fit for him, and that usually doesn't end well for the veteran or the employer. Both lose lots of money and time and end up unhappy.

So, as you're narrowing the field, it's important to understand the way each phase of the transition process messes with your head and your emotions. Are you still nervous about the transition? Have you gotten to the point where you are starting to become a little overconfident and maybe arrogant? As you get closer to the end, do you feel the pressure of your "out date" makes you want to accept the first offer that pays the bills? These things will sometimes color your view of your value system and what you really want to do.

So, as you're evaluating where you are, take into account the emotions that you're feeling and understand they are natural. Most people only make one military-to-civilian transition in their lives. It's not expected that you're going to know everything. What is important and what should be expected is that you're making well-considered decisions.

We strongly advocate seeking out transition assistance and a guide, someone who can bring you through that process. If you're working with Absolutely American as part of your transition plan, every single interview that you have will be debriefed with us to evaluate it against your criteria and determine how you and the people who are important to your decision process are feeling.

As much as possible, you want to have someone who can walk through that with you and have a very intelligent and well-considered conversation about it, because you are literally betting your paycheck and your lifestyle on the decisions you're making for this.

The last point about narrowing the field is to remember that few candidates will reach their goal in the first step. Frequently, to get to the goal that you want requires an interim step or two. Thinking more broadly, even your active-duty service has been an interim step toward where you're going.

This ties back to why people join the service in the first place. The mistake is to see the transition point as the end state. It's not. It's all part of the continuum of your career that's pointing toward where you really want to go. Ask yourself, *What am I doing? Where am I on the road to my goal? Will this opportunity take me closer to where I want to go and make me more valuable to the employer I'm currently working for?* If you do that consistently, you will find that, eventually, you will get to where you want to go.

Chapter 9
Offers

We've lived by the mantra that nobody has ever turned down a job because they were offered too much. Believe it or not, we've actually tested that theory a couple of times. A number of years ago, we were working with a young Army officer who had stepped out of active duty the year prior and was relocating to the Northeast to support her husband's attendance at a prominent university for his master's degree.

As part of the process, we were working with this young woman to make her first major move into the civilian career space. We had presented her with an opportunity for a leadership role in that area that really seemed to connect well with her and fit what she wanted to do with her life.

When we got to our final debriefing and were talking about the level at which she would accept the position, she gave me a very reasonable salary that it would take for her to accept the offer, and we felt that the employer would have no problem meeting it. As a matter of fact, we actually expected the offer to come in slightly above what it was that she had said, perhaps 10% or 20% above what it is that she had given us.

When the offer finally came in, it ended up being twice as much as what she had wanted for the position. When we presented it to her, she literally almost fell out of her chair, but you can rest assured that she did, in fact, accept the offer.

Obviously, these types of stories don't happen every day. As a matter of fact, in a lot of cases, especially now, the wiggle room in an offer process will be much tighter. Competition for good people has driven most employers to make the best offer they can up front, but the first thing a transitioning military veteran needs to understand about your first offer coming off of active duty is that their offer is not going to be based upon what you made on active duty.

That's not their primary area of concern. Certainly, they understand that what people want and what they have made in the past is an important part of the process, and there is a recognition that trying to be at or above where a person was is one of the criteria for determining an appropriate candidate. But it's important to understand why we're paid what we are for active duty.

The History of Military Pay Rates

First of all, the government expects us to perform any and all the requirements that go with active duty. The government has no obligation to pay us any overtime, and they don't. It doesn't matter that the military works us more than 40 hours per week, especially when we're deployed or out in the field or on a ship at sea.

Part of the military pay that we receive is based on the arduous nature of the work we do. It's also based upon what the services believe it needs to pay to attract and retain people to serve. We understand that a great deal of sacrifice goes with military service, including time away from home and the risk of life and limb. Because of the nature of military service, we don't have to pay taxes on a portion of our income, and that's certainly a benefit. It's also very nice to have our housing and food costs covered as one of those tax-free benefits.

Offers

When we became subject to the Uniform Code of Military Justice, the government took on the responsibility of housing us. In days gone by, the military would put all service members in base housing. But as the services expanded, especially after World War II, to maintain a large standing military again, the practicality of feeding all military members in the chow hall and housing everybody on base became impractical. So, the military had to rely on off-base housing to meet all the needs of the services.

That's why we're paid those things and why they are non-taxable benefits, as we refer to them now. That is an important thing for us to understand whenever we're making the move. The reason why those things are as they are is because, when you're on active duty, you are compelled to live in a particular area and have to be in proximity to where your unit is located.

The history of military pay over the ages has changed quite a bit due to a number of different economic and government-based decisions. Certainly, during the ramp-up in the 1980s, as we were building in the Ronald Reagan era toward a large standing military to face the threat of communism from the Soviet Union, upward salary pressure increased after the lower-pay era of the end of the Vietnam War and the inflation of the 1970s. Once the Cold War ended with the destruction of the Berlin Wall and the dissolution of the Soviet Union, military pay remained relatively flat through the 1990s, with very low pay raises for military members and no significant adjustments to how things were done.

That all changed in the early 2000s. If you're serving on active duty today, you probably do not realize that for the first time in nearly 40 years, the military adopted broad pay raises across mid-grade officer and enlisted ranks. From January 1, 2000, to January 1, 2002, the services implemented pay raises every 6 months, and reenlistment and retention bonuses expanded across the services. Note that this all started before 9/11.

Those pay raise programs arose due to the economic boom caused by the expansion of broadband internet and telecommunications and the investment into changing computers and software associated with Y2K (Year 2000), which led to massive resignations and exits of mid-grade officers and the enlisted from the services. In fact, at that particular time, one of the primary drivers to leaving active duty was that there was significantly more money to be made outside of the service than in it.

And in the early 2000s, the government recognized that the pay for this critical demographic inside of the military required a significant adjustment, so they made those changes. Very shortly after, of course, 9/11 happened, beginning the long war in the Middle East that has only recently come to a close.

During the period of war, to draw people in and retain them, the services forced additional 4% to 5% pay raises and increases in military benefits. The result of this was that, by 2008, the Military Times reported that the pay gap in comparison to civilian peers had been completely erased, and in the years following, the rate of growth in military pay far exceeded the increases found amongst civilian employers.

That has become problematic for military members who are leaving active duty. If you have a highly pursued skill set, come from a technical specialty, or are an officer performing in a role that was very relatable and desirable to civilian employers, your pay rate as a civilian will probably be a fairly even move across, and you may even be able to get a slight increase. But if you're coming from a specialty where the equivalent pay as a civilian is significantly less, it has become much harder to find compensation that compares favorably with what you are accustomed to outside of government contracting roles in your military specialty.

When it comes to veterans' employment in the civilian sector, this is a topic that is much less discussed than it needs to be, but it's a reality that service members face. This has led to a great growth in interest among service members to pursue employment with the federal government or government contractors because the government has indexed its pay to the military, making federal employment particularly attractive.

At the time of this writing (late 2024), Congress has proposed a nearly 15% increase in pay for junior enlisted to help encourage enlistments.

Knowing where it is that you are in this process and how it is that these things apply to your value system is very important to understand, because not having your budget aligned the right way or being unprepared for the realities of your situation can harm your ability to maintain financial viability as you move into the civilian workforce. It is important to understand that just because you made a certain amount while on active duty, it doesn't mean that other employers will pay you the same rate. This is the essence of our Military Free Agent™ training program: Creating competition for your services helps you achieve the best possible offers that you can get and not what you made on active duty.

How Civilian Pay Ranges Are Created and Maintained

Now that we're at the offer phase, we need to understand the concept of pay ranges. Where do they come from? How are they created? What's their purpose?

Employers create pay ranges to have some system for what fairness looks like for the people they hire to do a particular job.

It would be highly improper and ineffective as a recruiting strategy to pay everybody exactly the same thing for exactly the same job based only on one set of criteria, which would be the job itself. People bring all kinds of different skills to the table that can add value to a company. Therefore, companies have to be able to alter compensation for people as a result of the additional qualifications or character qualities that they bring that are relevant to the role at hand or to the future role for which they might be considered.

The other reason for pay ranges is that they create a system by which an employer is able to offer increases in pay for the same job to either offset cost-of-living increases or for continued retention inside of the organization without having to make a major change. Many states now require employers to post the entire pay range for a particular position.

Military members should not feel that being hired at the top of the range is frequently realistic. To explain this using the military model, it would be like looking at the military pay chart and the pay range for an E-5 and saying, "Well, I want the pay that's on the 'E-5 over 20' portion of the chart." Clearly, that's not realistic. To get that, you have to have 20 years of experience.

Civilians have the same mindset. The difference is that it's not all based upon time and service, but to a certain extent, it is based on where it is that you come into a company. Most organizations hiring from the outside pay somewhere between the zero percentile, or the entry-level rate, to the 50th percentile, or the "midpoint". So, it is usually the lower half of the range. This is important to understand for a couple of reasons.

One of the things that you're going to encounter as a civilian is the concept of "salary equity," which means that employers do not want to hire people from the outside at rates that are significantly higher than what they pay the people who are currently working for them. This would cause all kinds of problems, and it does during periods when it becomes necessary. Even though compensation in civilian organizations is usually considered a private matter between the employer and the employee, the reality is that people find out what their coworkers are making, and they judge the company's opinion of them based on the difference.

So, you should understand that it would be very unrealistic to believe that you're going to be hired from the outside for your first job at the top end of that salary range. It would be more reasonable to look at the 25th percentile. In other words, take the midpoint and the entry-level rate and expect something halfway between them.

But if you don't know what the range is—and this is still frequently the case—the important question to ask is: How valuable is this position to you? What do you believe is the lowest amount of money that you could take to do that job? This is the internal part of the process. We refer to this as the "money-aside" question. The first question is, "Money aside, are you interested in this position enough to want to take the job?" If the answer is yes, our follow-up question is, then what would be the minimum amount of money that you believe you can do this job for in the long term?

One of my very close friends is a West Point graduate who is now a division president for a $2.7 billion, multi-national, publicly traded company. A number of years ago, he shared with me his philosophy for accepting offers and how it has helped him create success. What he told me is that he always accepted the offer the company gave him as long as the opportunity was right and he could live on the amount offered. In other words, if the role fit what he was looking for and what he could do and filling the position for two years would make him more valuable to his current employer and other employers, he would accept it as offered.

His approach, especially if he believed that the offer was below what he wanted, was to inform the employer that he intended to perform at such a level that when he came back to do his one-year review, there would be a serious discussion about adjusting his pay upward. Then, he would accept the offer as presented based on the employer's willingness to accept that condition.

Over the past 15 years of his career, he has multiplied his compensation rate by almost 15 times since coming off of active duty. It's been extraordinary to watch. By outperforming what was expected of him, he demonstrated his value. Consider this in the context of what we talked about earlier: the sign that says *"Help Wanted"* on one side and *"Headaches Need Not Apply"* on the other.

The strategy that this individual used was based on the fact that he was willing to prove his worth in each of the situations he went into. He understood how important it is for his employer to have confidence in his value. If the company he worked for was not willing to adjust to that, then his philosophy was: "I will look, and I will find somebody else who will pay me what I'm worth." He believed in himself as his own job security.

Needless to say, different jobs have different limits. The higher you are in an organization, the more negotiating leverage you have. Most people in hourly positions will not be able to negotiate large increases in hourly rates in their second year. Sometimes, the company is willing to take a chance on giving someone a shot at the role but is not comfortable risking a lot of money on them.

The most important part in all of this in terms of life as a free agent is that your job security for the remainder of your career has nothing to do with your employer. Your job security is solely derived from your willingness to continue to learn and grow in your career, and your willingness to walk away from employment situations that are not win-win balanced makes employers take you more seriously. This is the difference between having a career and having a job. If having a job is your focus, it creates a sense that it's the employer who is responsible for your security. Why give control of your security to someone else?

When your focus is on growth in your career, then the employer is certainly part of that equation, but most of the responsibility rests on you and is built around your willingness to make changes. Everybody has different levels of tolerance for this. The process that I have just described here is for the person whose interest is in growth and being able to take on more. Their interest is to be able to become more in their vocation tomorrow than they do today. That may not be in your value system. Your value system may be to work in this role, have some growth, and stay with one employer for the rest of your career.

There's nothing wrong with that. That's very noble, and many employers will appreciate it, and it's also the reason that many people take government jobs. But the problem with that philosophy is the part where you take your hands off of control of your own fate and become resistant to the idea that someplace else might be better. Remember the words of Jim Collins: Good becomes the enemy of great. But again, these are personal choices. My objective is to empower you by showing you how to maximize your value as a free agent and minimize the likelihood of significant unemployment in your life. It is not to tell you how to live your life.

Common Elements of Employment Offers

What are the different elements of the offer? Well, certainly, the base rate of compensation is part of that. Some positions have some type of eligibility for overtime. Others can have incentives in their bonuses based upon the attainment of different goals. That could be a combination of individual and organizational goals.

Some have unplanned incentives. They're provided to you if you accomplish something significant. And some companies have no incentive type of compensation at all. These are all things that can fit into the base level of compensation. Your ability to have these different things is often based on the class you're being hired at.

There are three primary classes of employees inside a company, and each has different rules as they relate to labor laws and what can be negotiated. The three categories are labor, management, and executive. Most of the labor laws are built around establishing fairness across the three categories and protecting the labor category, which is viewed as the most vulnerable due to having the least influence.

For example, when it comes to health care benefits, the government is very meticulous about ensuring that what is made available to executives and management is very similar to what is offered to labor employees. These laws exist to ensure fairness in the workplace and prevent people from being taken advantage of by people who have more power. So, when the government is involved in certain portions of compensation, such as health care benefits, there are going to be significant restrictions on what can be negotiated.

Another item that can have great degrees of variability, depending on the organization, is paid time off. There are many ways that paid time off can be offered. It can be a combination of sick days, vacation days, personal days, floating holidays, paid holidays, and a lot of other things.

Some companies are open to the idea of paid time off as being negotiable. Other employers will have very strict standards that are purely based on the amount of time it is that you have worked for an organization and won't change it either way. What you have to remember is that employers have multiple objectives when it comes to the compensation and benefits packages that they offer. The first part is in the nature of how it's described in the first place: "compensation."

A person performs duties in a particular role, and the company has a responsibility to compensate them for the work that they've done. The government has set minimums that companies have to pay and also has regulations related to the amount of time that an employee can be required to work. When it's over 40 hours a week, some additional compensation or some differences in compensation have to exist for this. But the idea behind this is that people who do their work have to be compensated for what they do, whether they're labor, management, or executive.

The second part of compensation and benefits has to do with retention. The value of the benefits becomes a reason why people want to work for a company and don't want to leave. That is one of the reasons why some companies can be very strict with their paid-time-off policies. It encourages them to remain with the current organization, as leaving would cause them to lose that particular benefit.

Some companies offer retention bonuses to keep people, such as a cash incentive for someone willing to stay for a period of time. Why? Because if the person leaves, the company will incur a significant cost to find another person to fill that particular role. So, if a form of compensation can retain somebody at a lesser cost than would be spent finding a new person, it can be very smart of those leaders to use it. Annual raises are another incentive for retaining people.

That's one of the reasons, discussed earlier, why you have a pay range. The challenge, though, is that, ultimately, you may very well reach the top of the pay band. For example, while I was still on active duty, my wife worked in the medical field in Virginia. The company that she worked for paid her near the top end of their pay range for the position she was hired to fill. That was because we were moving from an area that had a higher pay rate for the work she did.

After three or four years of working there, the owner of the company came to her and said, "We're giving you this raise, but it takes you to the top of our pay scale. You need to understand that going forward, we're not going to be able to increase your compensation further."

My wife decided that that wasn't acceptable and that she would look for another position. She was working in an outpatient facility where she had a great degree of freedom in her work and good staff that she worked with, but she was dissatisfied by the idea that she would never be able to make any more. So, she accepted a position with another organization that offered her a significantly higher compensation rate. In the end, she found that the company that she went to had a culture that was so caustic that she decided she could not work there, so she returned to the company she had been working for and stayed there until we moved out of the area.

Pay Attention to Company Culture

My wife felt that the benefits of the culture, the freedoms she had, and the type of work she was doing were worth more than money. This leads us to the next part of compensation and benefits that isn't monetary, and that is a company's culture. Part of the process of due diligence in looking at a company is to understand its culture. You're going to spend less time dealing with your base pay rate and your benefits than you will with your job and the people.

When people and jobs come together, it forms a culture, which is how people work together to do a particular function inside of an organization, and it becomes the identity for how an organization works. One of the things most overlooked by veterans coming off of active duty is the culture of the company that they're going to work for. In most cases, they never look at it and don't even know what to look for. In the military, they've had to deal with many different cultures—some of them good, some of them not so good.

Usually, a military member will have had experiences with both in their career. Part of our process in teaching them about the transition is that one of the benefits of being a civilian and having the ability to make change is being able to leave a culture that is not suitable for who you are and what you want. We try to make sure that they consider the company's culture and make that part of the due diligence they do before accepting a position.

Six months after your transition is not the right time to find out that you landed in a company with a culture that doesn't fit you. You may have been successful in your transition and finding a job, but the misery that comes from being in a company whose culture doesn't fit what you're looking for or doesn't give you any enthusiasm for the work you do or the people you do it with is ultimately deleterious to your mental health.

That's why we see so many military members make changes in their employment within the first year. As a matter of fact, it's been our experience that between 40% and 60% of military members who leave active duty and find a position on their own will leave that company within the first year. That's not good. That's not healthy. The key to success is knowing what you want and making a plan to get it, but let's get back to the most important thing: negotiating your pay range.

Negotiating Your Offer

Currently, there are very few organizations that are going to try to attract their help at the lowest possible rate; that is actually a short-sighted view of the problem. Ultimately, the goal for a company is not just to get somebody to say yes and come in at a low rate and, therefore, keep a person underpaid for the remainder of their career. No, companies understand that not only are we trying to be smart with our money and acquire people at levels of compensation that are reasonable and cost-effective, but at the end of this process, we have to be able to compensate our employees at a level that will keep them with the organization. The cost of turnover is high.

We have seen significant change in industries across the board as the job market has remained strong for 15 years, with the exception of the short period during the COVID shutdowns. That strength is unprecedented in American history and has completely changed the way employers recruit, pay, and retain people. Companies who have tried to lowball the people who come to work for them find themselves with no employees and wasting tens of thousands of dollars trying to hire new ones.

You've probably heard the saying "penny-wise and pound-foolish." That's what a lowball offer is—penny-wise and pound-foolish. Thankfully, the market has driven employers across the board to make better decisions in terms of their offers. So, employers will usually make the candidate they have in front of them the best offer they can justify, which means it will be based on a combination of the candidate's qualifications and experience and how much other people in the organization are making who are doing the same job.

It has been our experience that offers for most military members are only going to be negotiable up to a maximum of 5% more than what has been given. Most of the time in today's market, the negotiability of offers is much less. We're finding that from zero to 2% negotiability is most common in the current market because employers understand that they need to put their best offer on the table first. Otherwise, the person they're interested in hiring is going to be turned off, and the employer will have to start the process all over again.

If you remember this from the balance-of-power discussion in previous chapters, by the time we reach the offer phase, the balance of power has shifted slightly in the candidate's favor. The employer has said "yes," but it is the job seeker who is in control of whether it is a yes or no in the end—and that is very important. This is where there is an occupational hazard for the candidate who needs to understand that their power here has limits. One of the worst things a candidate can do at this point is reply with a counteroffer that is too far above what has been presented.

For example, if you have been offered $30 an hour for your position and you come back to the company and say, "I want $36 an hour," most of the time in this market, the client will withdraw the offer because they will feel that you are looking for more they can afford and will leave the moment you find it with another employer. Again, if you were to switch positions with the employer and it was your money that you were spending, you would probably feel the same way.

What About Health Care?

One of the things we see frequently is the persistent myth about being able to negotiate away health care benefits from employers. Prior to the passage of the Affordable Care Act, the negotiation of health care as a benefit was absolutely a non-starter. But it has been talked about in military circles for as long as I have been in the transition business.

What limited this was the 1993 law passed by the Bill Clinton administration called HIPAA, or the Health Insurance Portability and Accountability Act. Most people think that this law only governs the privacy of medical information. However, HIPAA also made employer-provided benefits mandatory for all employees at almost any time (life-changing circumstance) regardless of any negotiations between employer and employee.

The idea behind that, particularly with military members who have ongoing benefits from the VA or the services themselves, was that if you didn't need a company's health care benefit, other benefits could perhaps be added as a way to adjust for that. Ultimately, employers had to decline that because at any time or during any open enrollment period, you would have to be given the company's health care benefits for it to be able to maintain the tax deduction for health care premiums. This was a shock to many people who were on a military pension, but the military "pension" is actually more of a "retainer" than it is a pension.

The rigidity of this system has actually loosened up a bit with the passage of the Affordable Care Act and the prevalence of healthcare savings accounts and high-deductible insurance plans. You will find that some companies now offer some type of cash incentive for people who choose not to enroll in a company's health plan because they have health care options from another source.

Those cases still permit the person to obtain the company's provided health care plan during the enrollment periods or major life changes, but it's open to everybody, and it can now be offered as a stipend and paid that way. Companies that offer this will declare it in a benefits package if that is offered. Ultimately, the thing to know for the military member is that this is a structured benefit alternative and not something that you can negotiate as far as your pay in today's world.

So, in summary, the most important part of the offer package is not what the company offers you but what you believe is the level of compensation and benefits necessary for you to take this step in your career. Certainly, the offer is important. The compensation is the reason why we do the work that we do. But this transition and this offer that you're accepting is not the end of the journey. It's just the beginning that you get to control.

Short of any obligations that you're being given, you're going to be able to make a change anytime you want. The Department of Labor expects that most employed adults in America will make six major career changes over their employment lifetime. Understanding how to navigate this process will help you acquire employment that not only meets your work and environmental goals but also satisfies your financial goals.

One last point on offers: Employer-paid relocation benefits have a great deal of variability and can still be negotiated in different ways. The first thing to remember is that many jobs open to transitioning veterans will not have a relocation benefit option, but some will. Even with companies that have structured plans for this, if they offer relocation assistance, there are often degrees of variability. Many companies will use distance and family size as a way to calculate this, but there's no regulatory requirement to have a standardized relocation package. Such benefits can vary greatly, from fully funded relocation packages to something as simple as a cash relocation benefit to assist with the expenses. As I mentioned, not every company offers this at all, and even fewer will offer it for the kinds of jobs transitioning military members can fill.

In fact, the number is actually very small, but it is something that can be very helpful and is used to attract people to a company. What you need to know is that any type of upfront money, especially in a relocation benefit, will typically involve some period of obligated service that is tied to the cost of the relocation package. Generally speaking, most employers do not require more than a two-year obligation of service for a relocation, and this is generally reduced on a quarterly basis. In other words, for each quarter that you remained employed, your liability for the relocation benefit would decrease by 12.5%.

This can also be true of new-hire bonuses, or what's commonly referred to as sign-on bonuses. Some sign-on bonuses have obligated service requirements. Nowadays, it's becoming more popular to refer to them as retention bonuses that are advanced rather than sign-on bonuses, which implies that the value of the bonus has not been earned until the retention period has been fulfilled. We have a number of clients that do offer these types of bonuses at times and now refer to them as retention bonuses, and the money the person receives up front is actually a draw against future compensation earned at the end of the period of retained employment.

The structure of those agreements indicates that the benefit is technically only earned after you have crossed a particular length of employment. Therefore, you would be fully obligated to pay it back should you choose to leave short of that goal, which is fair.

Are there times when companies will break pay rates without consideration of internal equity?

Yes, there are severe shortages in the industry. What you will find with most employers is that, periodically, an entire pay band has to be adjusted because market forces are driving it up. Of course, the same thing happens in the other direction, where market forces can drive the particular need for a type of position down to zero, and those positions remain stagnant, and then, frequently, they can be eliminated.

But when we see these types of things happen, it usually occurs during a growth economy. We have seen that these ranges will exist for about seven to ten years. What happens toward the end of that progression is that the market drives the cost of bringing new talent into this industry, which is already hiring people from outside of their specialties to meet their needs and now has to hire people with relatable skill sets, higher than the people who have already been trained and have experience in the business.

The company's need to find people justifies hiring somebody at levels above the salary equity. The problem, of course, is that they have to resist moving the entire organization to that level because, from a financial standpoint, they have not made the sales necessary to move the entire range up. For a period of approximately three years, the new hire rates exceed the existing employee rates to the point where enough of the existing employees have discovered the change, have not been able to make changes to their own compensation, and have started to leave at a rate that now is causing additional problems for the organization. At that point, the company realizes that a change is necessary, and they move the entire pay scale up. We have seen this happen three different times over the past 25 years with a number of organizations that we have worked with.

This is understandable. It's just how people think. As a perfect example of this, if you owned a house before the interest rates changed, your idea of what is needed to live is based on the interest rates you enjoyed when they were at historic lows. But the new person trying to come to your market doesn't get to buy in at the lower prices and interest rates. So, it may take more money to bring in this person.

There are many reasons why things tend to follow a seven-year cycle of change. For example, on average, people in America tend to relocate and sell their houses every seven years. Things like that become the main drivers of change.

Summary:

1. Remember, money aside, do you really want the job being offered? If the answer is yes and it's your best choice of job, people and company, then very frequently, it will be your best offer for you to accept and will accelerate your growth. Many highly compensated positions have unpleasant downsides that have to be covered with compensation to get people to accept. As we showed, even our government believes that military service is a job that requires higher compensation than other offerings to get people to join.

2. Company culture is the single most important item that military veterans fail to evaluate and the most frequent reason for changing jobs within the first year. Make sure it fits you.

3. Offers have lots of components that affect your net income and lifestyle. Be sure to look at them carefully because little is standard. No two employers are the same. Starting base rate negotiations will not likely be much in today's economy, but future reviews can be. Also, few benefits options can be negotiated but may have built-in flexibility

and incentives. Health care benefits are highly regulated. If relocation or some kind of up-front one-time payment is offered in the role you want, it is usually the most negotiable option.

4. Salary equity will always be a big challenge for employers. Whether you are trying to find a position or are working for a company facing headwinds to attract good employees, you may experience both sides of that problem at different times in your career. Using your value systems as a guide in both saying "yes' to the offer or in making the decision to stay or go will always help you make the best decision for you.

5. During the civilian transition, you are not likely to be hired at more than the midpoint of the salary range unless your experience from active duty matches the role very closely. Even then, being hired too high on the scale may provide challenges for future growth in the pay band.

Chapter 10
The Government Sector

The government sector is just like the military, except that it's nothing like the military. When it comes to federal employment, veterans are frequently attracted to working for the federal government because military service can be relevant and the benefits there are real, as the federal government has a system to give veterans credit from their time in service toward a federal retirement, which can be highly attractive.

The United States government is actually our nation's longest employer. Dozens of departments are found nationwide and around the world, and technically, you're working for the same employer, anyway. One of the key benefits to military service is that drawing a military pension and working for the federal government is not considered "double dipping" by the laws as they are currently written, which gives a military "retiree" an earnings advantage with the federal government.

Federal government employment also has a slightly different process when it comes to hiring. Everything that we have talked about previously remains valid. There is still a need that causes a position to become available. However, the federal government has certain requirements governed by law and policies that make getting hired by the federal government a little different and a bit more bureaucratic, as you might expect.

First among these is the federal resume. Though not required for every federal position or agency, it is for the vast majority. A number of organizations can help with building a federal resume, and we have included examples at CorporateGray.com. Most types of training on resume writing are outside the scope of this book, but you can certainly follow the links that we've provided and some of the reference materials that you can find on the Corporate Gray website. Beyond this, the process of conducting your interview is still going to be the same. A government interview will usually be much more structured than what you'll find in most of the commercial sector, but the kinds of answers you provide are going to create the exact same impression.

How do I know that?

In 2002, my youngest brother was coming out of a technical school in electronics, and as he was preparing to graduate, the U.S. Department of State was conducting interviews at his campus.

He was highly interested in the opportunity presented to him, and I helped him build a resume. The nature of the role did not require his resume to be in federal resume format at that time.

In preparing my brother for the interview, I had taken him through all the principles included in this book: how to ask questions, how to position himself, and how to emphasize certain aspects of his school and experience that he most enjoyed doing. When the final count was in, the State Department had interviewed more than 500 people in the Pittsburgh area during their campus sweep, and my brother was one of only five who was actually hired into that department—and the only one with just an associate's degree.

So, the things we teach work even with the federal government, because there is still a problem that demands a people solution, and people are all influenced the same way. Of course, the federal government is not the only level of government that hires service members.

The Government Sector

State governments around the country also have an interest in hiring veterans, and many of those give credit for military service as part of their retirement plans.

Of course, state governments are much tighter geographically (especially the smaller ones), and they are still very large employers in every state—if not the largest—and have lots of departments. Like federal government employment, state government employment is still a public service. However, the variance around state government employment is so broad that it would be difficult for us to cover the subject in this particular book.

However, the primary focus, again, of solving the problem that the organization has is always appropriate. Your state resources will provide the information on what you need as far as a resume, cover letter, and supporting your experience are concerned, as each state will have its own requirements.

Local governments are similar in this, except, of course, the focus there is impacting the local community, specifically where it is that you live. Of course, they vary greatly in size. When you look at the fact that New York City is a local government entity but is larger than all but the five most populous states, you can understand that some of these large local government offices are going to behave more like state governments than local ones.

Of course, local governments can go down to very small towns and townships where you may have very little paid government. As a result, they can vary greatly in terms of their pay and benefits. Often, a veteran who is looking to be geographically targeted will opt to work in a local government position for the benefit its location provides, along with the ability to continue to serve.

Working in local government can be a great long-term career. In many cases, they are feeders to larger government organizations and departments, particularly if your interest is in law enforcement. When it comes to working in the government, military service continues to be valuable, and if it's appealing to you, you should absolutely pursue it.

You can also find employment with government contractors. The kinds of things that government contractors hire veterans to do are typically related to the military and its support. Most will not take a job with Lockheed Martin and work in a plant that actually builds F-35 aircraft. Large government contractors do much more than just make the products that the military has. Significant portions of their organizations are dedicated to providing support in the form of staffing contracts to do certain parts of military work that the government has decided it wants to outsource.

Maintenance contracts are common. Many military bases have a significant number of contractors who perform maintenance on military aircraft or military ships and submarines, doing the work as civilians. These military veteran contract workers perform repairs and upgrades at a deeper level than what is typically done in the military, but now they are doing it for the company that makes that particular product for the military. This is one of the things that makes this type of work extremely appealing to military members. The skills that you have developed from active duty have a direct application to the work that you would be doing for the government contractor and in a location you're already familiar with, maybe even inside of the same shop.

Beyond this, there are other types of roles that government contractors like veterans for, and many of them have to do with different types of program management. Junior military officers, in particular, are frequently hired to manage programs or do analysis work related to what they did on active duty or relative to the service as a whole, if it's more of a general nature. This allows the junior officer or the senior non-commissioned officer to be able to make this transition, working in a space with which they are already familiar and where they speak the language.

The other benefit to government contractor experience is that these types of contracts are set up to hire military personnel based on the level of veteran that they are looking to hire. For example, maintenance contracts where an experienced E-5 performs most maintenance are priced relative to the government's cost for supporting an E-5. So, when you make the move to the contractor space, you're likely going to be paid the full equivalent salary and a portion of what the government's overhead costs were.

As a result of this, many military members can move into these types of roles and actually enjoy a pay raise over what they were doing on active duty though they won't have the same kinds of benefits. The same is true with program manager and analyst positions, which may be targeted for 03–04 level military members. Therefore, when contractors hire those people, it is based on that pay rate.

The same thing applies to senior officers and the highest ranks among the enlisted. Such opportunities allow you to have a nice increase in compensation while supporting the mission without having to wear the uniform. You can leverage your military skills and the functions that you've enjoyed and build depth on the systems and programs you need to know. Frequently, there's no need to relocate to the base or area that you want because, typically, you're already there.

All of this probably sounds really good. What's the downside of all of this? As we know, for every upside, there is always a downside.

Current Events Note

DOGE. Since I began writing this book, The Department of Government Efficiency or "DOGE" has become widely known across the country, especially inside of the federal government. In pursuit of cost reductions to ease pressure on the national debt, the Trump Administration has been reducing the number of federal government employees and terminating unnecessary contracts with the federal government. At the time of this writing, DOGE has only been working for three weeks, and the President has placed a hiring freeze on nonessential hiring within the federal government. The Department of Defense and Veterans Administration have not been fully examined, but the effects of cuts at USAID and the Department of Education indicate that we may be entering a period of lean hiring within the federal government and its contractors. Only time will tell what will happen.

The Contract Cycle

When I was leaving active duty, I was a part of the Navy's Operational Test and Evaluation Force. The role of that organization was to do operational testing and evaluation to determine if systems were both "effective and suitable" for fleet release. As you can imagine, a big portion of this had us dealing with vast numbers of government contractors on all the different types of work we did. At the time that I was making my transition, the primary contractor that we were working with was on a trip with me to do an operational test. I asked him about moving into government contracting, and what he told me has stuck with me ever since.

He said that the world of government contracting is for the person who is looking for stability and not necessarily growth. Yes, the contracting world can create a nice income coming right off of active duty, but each of the contracts has a particular duration. Frequently, labor support contracts run on a five-year cycle, and during the five years it is that you would be working for a contractor, your pay will increase slightly each year, but at the end of the contract, there is no guarantee that the employer will be able to retain the contract again. Another company could very well underbid them, and they frequently do.

If that's the case, the employer changes, and technically, you lose your position with the organization that is currently employing you. You could be offered a position by the organization that's taking over, but usually, it is at a lower pay rate, as the other organization got the contract because they were able to underbid the contract that the previous organization had.

This process continues over time. Of course, the other part of this is that certain contracts only have a certain amount of life in them. If the system or program that you're working on becomes obsolete, you'll have to find a related position based on contracts available at the time. Many people look at the downsides and feel that they are more than compensated by the upside of being able to continue to serve as they have and with the lifestyle that they're able to enjoy.

If you are comfortable with this employment model, working in government contracts may be a good fit for you.

Using Your GI Bill for Full-Time School

The single greatest benefit ever provided to military members who are currently serving is the current GI Bill. As someone who served prior to 9/11, it's easy to be envious of the benefits that have been provided to military people today. Among these is the GI Bill.

Prior to this current GI Bill, enlisted military members had to contribute during the first year of their enlistment to get the benefit. You could only use it for a limited time, and it paid far less than what the current GI Bill does for schooling and benefits. If you were an officer and had your undergraduate degree paid for by the military, either through an ROTC scholarship, one of the service academies, or some other type of college program, you were not eligible for the GI Bill at all and left the military with no benefit whatsoever.

Full tuition payment up to a certain amount plus housing allowance for the time that you are attending school makes for a very attractive benefit. The idea of going to either undergraduate or postgraduate studies using this benefit is a strong draw for many people coming off of active duty—and it should be. This is a benefit that is well worth using. What's important to know about the GI Bill benefit is how you use it.

What I'm referring to by this is how you manage to support yourself while you are attending school. One of the things that we frequently hear when suggesting that someone use the GI Bill benefit is that little thought has been given to how the veteran is going to support themselves while they're going to school. Though the housing allowance is a terrific benefit and will largely cover the cost of living someplace while you attend school, it represents less than a third of what you are making on active duty.

Among the challenges that we discussed from coming off of active duty, almost all military members join the service with almost no personal and financial responsibilities, and many of them—almost all of them—leave with significant amounts of financial responsibilities. So, when examining the challenge of going to school full-time and living off of the GI bill, the prospect of having to find other types of income to fill the gap between what the military offers to you in terms of your housing allowance and what it is that you can make up for in other ways can cost you a lot of time.

We frequently hear things like, "Well, I'll attend school full-time and get a part-time job at a bar." When you start putting pen to paper on that, the one thing that ends full-time is the "part-time" job, and school gets pushed to the side. For most veterans, above everything else, is the pressure to find more money than what the GI Bill will cover, and I literally just dealt with another veteran today who tried this and found it doesn't work when trying to live on your own.

So, before you decide to take the plunge into full-time school, I encourage you to really look at your budget and make sure that you have figured out a way to live that doesn't require you to work full-time while you try to go to school full-time. That can be a self-defeating process. Undergraduate education studies, in particular, require a great deal of your time, and investing in them to get the value out of what it is that you plan is important.

That's the word of caution. The great news, again, is the value of the benefit. Once you have figured out a plan for how you're going to maintain your lifestyle while you go to school, there are an incredible number of options that are developing and have existed at both universities and many types of non-traditional education institutions that are available to you. For example, at Corporate Gray, we work with many non-traditional educational opportunities that teach people everything from electronics to cybersecurity and even how to do sales in high-tech systems. Many of these programs are eligible for use with your GI Bill.

Growing a career in the modern world demands a commitment to continuing to educate yourself. This can be a great way for you to do that, using that benefit without having the long-term requirements of actually having to attain a degree. However, most will look at this for the purpose of getting a four-year or postgraduate degree.

The Student Veterans of America (studentveterans.org) is one of the best organizations that veterans can connect with once the decision to enroll in a college or university has been made. With Congress paying increasingly more attention to the use of the GI Bill and the success of veterans in completing their programs of study, organizations such as the Student Veterans of America have become increasingly more important. SVA provides an environment where veterans can come together and take advantage of the shared experience that comes from military service and the challenges faced as part of pursuing an education.

It can also help with networking and developing job opportunities when coming out of your educational program. But mostly, the value of this has to do with dealing with the unique aspects of attending school full-time as an adult, as opposed to someone who is coming right out of high school, and they'd be highly recommended to do that.

Many colleges and universities have special departments specifically for veterans that will help guide you through the process. However, when selecting where it is that you want to go to school, consider starting your process with the end in mind. Then, work that process backward to the program of study that will get you there.

The Crown Jewels of a College Education

It has been my experience that three elements of a degree program can act as career accelerators. What I mean by "career accelerators" is that these are elements of your university degree that will cause you to stand out and significantly move ahead of your peers after you graduate. I refer to these as the "crown jewels of a college education." Put all three of these in your resume, and your education experience will move you to the front of the line for almost any opportunity you want. The three crown jewels of education are:

1. **Quality of the School.** Schools that have been highly recognized for their quality of education stand out regardless of what is studied. This is no surprise. Everyone who chooses to attend college understands that certain schools have great reputations, which makes competition for admission fierce. That attracts opportunities for students while at school and in life.

2. **Quality of the Program of Study.** Even if the school itself is not widely recognized for the quality of its education across the board, programs of study that are recognized for their quality or even their challenge stand out to employers. Complete an electrical engineering degree from an accredited college or university, and the world will recognize that you stand out because electrical engineering degrees challenge you not only mentally but also physically. The quantity of work associated with this kind of degree and the difficulty of the program deter many from trying and more from finishing. Some programs are not as obvious to everyone. For example, though a business major at every university may not cause you to stand out to all employers, the undergraduate business programs at the University of Virginia have been highly recognized for their quality for decades, and competition to get into those programs is among the highest in the nation. Then, the work really begins. Bottom line: Choose a program that is recognized for its challenge and quality, and regardless of the school you

attend, you can expect that completing your program of study will open doors for you that are closed to others.

3. **Personal Excellence.** The truth is that your ability to capture the first two crown jewels of a college education will likely have been determined before you make your first application. Your past educational performance and standardized test scores will determine whether many of those options are open to you as a freshman in your undergraduate program or a first-year graduate student. But that's OK, because anyone can claim one of the crown jewels for their own, and that crown jewel starts with a personal decision. Once you decide to be excellent in your studies and commit to making whatever effort is required to succeed, you are on your way to claiming one of the three crown jewels for yourself. Completing an undergraduate or graduate degree is a grind, and consistently producing excellence over a long period of time will earn you the respect of employers—especially when added to your military service.

If you find yourself unable to capture one of my aforementioned "crown jewels," don't despair. All who choose to pursue advanced education take on a worthy enterprise, and the value of education is more than just a degree or even a major. It is the value of expanding your mind through the experience of learning.

When you pursue a degree, you choose to set other parts of your life aside to better yourself, whether you attend full-time or part-time, and even if you are using your GI Bill benefits, you are still paying for the privilege of the educational experience, so make the most of it. Earning your degree within the constraints of your life should be an accomplishment you are proud of. If you are attending school for the first time after high school, even finishing your first classes adds a great building block to the foundation of your future success. Regardless of what kind of education you pursue, continuous learning stands out as one of the keys to a happy and successful life.

Starting a Business

The correlation between military experience and success in franchising should not be a surprise. Franchises offer opportunities for people who have the initiative and drive to create their own success stories and control their lives by building a business. To support that, franchises provide opportunities through their training and systems, bringing people along to run successful businesses based on the pattern of previous businesses and how they've been run. In other words, franchises follow the military model. It's a training plan. It teaches people the system of how to do this by going the way of those before them.

That falls very nicely into the pattern of military behaviors that have been correlated with success in many different areas. Add to this a goal orientation and a desire for control of one's own fate, and it's no wonder that franchising can be a great opportunity for a military member to start a business—for themselves but not necessarily by themselves.

At Corporate Gray, we work with organizations that help connect military members to appropriate franchises. As part of that process, these organizations, at no cost to the veterans, help them understand the kinds of franchises that are available, where they would need to be located, how to obtain financing, and how to prepare themselves for a life working in the for-profit world.

Typically, one of the most difficult things that a military member faces if they have no exposure to business through family or other types of work is the idea of working in a for-profit entity where sales are a critical part of long-term viability. The great news is that franchising can take care of that. Of course, there is always the option to go it alone. Certainly, using guts and smarts, you can do that. However, the learning curve that is required to start something that you have not done before can sometimes lead to the kinds of catastrophic failures that end businesses and cost people a lot of money.

In the late 1990s, Robert Kiyosaki wrote a book called *Rich Dad Poor Dad*, the first in a series of books on how people can become wealthy. His second book in the series was titled *Rich Dad's CASH-FLOW Quadrant*. In it, he describes a model of the four different ways that people can legally make money.

The left side of the quadrant focuses on people investing time into something and getting money in return for it. The most common of these is what he calls the E quadrant, or the employee quadrant. When it comes to being an employee, regardless of what work it is that you do, even if you're highly commissioned, you're paid for your time. We know this because if you stop putting time into your enterprise, you will stop being paid the money.

The same is true for those who reside in the S quadrant, or self-employed quadrant. These are people who would consider themselves to be in business for themselves, but the role they're playing, the service they're providing, or the work they're doing still depends on their efforts. So, the self-employed person effectively owns their own job because, if they stop putting time into the enterprise, the money will stop as well.

What's significant about the other side of the CASHFLOW quadrant, the right side, is that most of the big money in the world is made through it. The right side of the quadrant describes duplication. Tops among these is what we call the B quadrant, or business quadrant. It's designed to duplicate the time of the business leader. The business leader creates a system of training and doing a particular part of an operation—or an entire operation—that can be taught to somebody else or given to another organization.

As a result, the business is able to propagate and grow without the owner constantly having to invest in it. By developing a system and the people within that system, the entrepreneur is able to take their relatively small amount of time and leverage it into a lot of time.

From that large quantity of time, they're able to derive income that doesn't come directly from their own work. The systematized business is able to produce far more than what the owner could produce on his own.

The last quadrant is called the I quadrant, and this is the investor quadrant. The role of the investor quadrant is to put money into something and get more money out of it on the back end. The connection between all of these is that, to create long-term wealth, the business owner uses the excess cash from things that are produced outside of their individual work and invests it in passive and portfolio-type assets that can grow over time. The idea behind this is that they're investing money that wasn't made by their own labor and are able to put it into something that will allow it to continue to grow further.

The importance of creating a systematized business cannot be overstated. Around the same time that Kiyosaki wrote his book, business guru Michael Gerber wrote one called *The E-Myth Revisited*, the "E-Myth" being the entrepreneurial myth. What the entrepreneurial myth refers to is doing the thing that you love and turning it into a business.

In the book, Gerber chronicles the story of a woman who loved making pies and was renowned in her area for how good they were. Her friends convinced her that she should start a business making pies, which she did, and she became fabulously successful at it. The demand for her product was very high. As a matter of fact, the demand became so high that she had to devote her entire life to making pies to keep the business going. Over time, she tried to hire people to help her out, but she could never get it done. Instead of owning her business, her business owned her.

Gerber consulted with this woman, and by teaching her how to systematize parts of her business, she was able to return to her love of making pies without them owning her life. The lesson from the story is the importance of creating something sustainable.

"Sustainable" is a term that's used quite a bit today, but ultimately, it refers to any type of system that can continue over time without significant peaks and valleys that might require additional inputs. So, if building a business is something you are interested in, we encourage you to choose one that will use the strengths of your military background. Having a system to set it up and another to operate it can help you duplicate your time in a way that makes your business sustainable—and franchising is a great way to do that.

Another opportunity that is unique to this particular part of history is acquiring an existing business. The baby boomer generation is the largest in American history, three times the size of Generation X and very similar in size to the millennial generation. Many of the baby boomers have created very successful businesses over time that have sustained them through thick and thin. One of the travesties of business today is that as this group retires and leaves the market, their self-employed and smaller B-quadrant businesses are being closed at an alarming rate.

This is happening not because the businesses aren't successful but because they could not find a buyer. My ownership of Corporate Gray is a perfect example of this. The founder of the organization and his wife ran Corporate Gray successfully for 30 years, but as they approached the end of their working lives, they began to look for options to find a potential buyer. They reached out to a number of people and organizations that they thought might be interested, but for one reason or another, none of them were interested.

When I first came into contact with the owner, he was planning to shut down this successful 30-year-old military transition resource business at the end of 2024 because he hadn't found anybody to buy it. This story is all too common today. I've witnessed it in areas where businesses are thriving, but as the owners age, there's nobody to take over once they're gone.

If you have an interest in taking over a business, the time that we are in offers a unique opportunity. Successful businesses that have been in play for a long time are actually much easier to finance than new ones. The business is already established and has immediate cash flow, employees, and operations. In many cases, these businesses also have owners who are motivated to continue the legacy they've been pouring their hearts and souls into for 15, 20, 30, 40, even 50 years.

In this market, established businesses are a bargain when they are available. Because of the number coming open on the market, many of them represent good opportunities for people to take over. When it comes to financing these businesses and making the acquisition, nearly every area around America has a local business development office designed to help entrepreneurs start businesses. These organizations can assist you in getting the funding needed to acquire, expand, or initiate enterprises that can create jobs and improve the American economy.

Of course, from the government's standpoint, the goal behind all of that is to continue to increase tax revenue for the things that it wants to do, but that's beside the point. The main idea is that resources exist to get these types of businesses. You would be buying a business where the owner knows what to do, is willing to help, can give you a very good deal on it, and can teach you everything you need to know to keep it going. So, if having your own business is important to you, here are links to a few resources where you can get more information on acquiring or starting one.

Conclusion

If there is one thing that I have learned about leadership through my time in the Navy and in leading veteran recruiting firms, it is the importance of setting clear expectations for what you want from your people. When people know what is expected of them, they take action with confidence, and when it comes to the vast majority of veterans, knowing what is expected almost always leads to results that are on target and on time. In my experience, not knowing what the people on the other side of the table expect is the single biggest impediment to securing employment when leaving active duty. Therefore, helping you understand what employers want from you and what they are thinking has been my goal in writing this book.

I could have written this book like a field manual and reduced it to just the facts with lots of reference materials. Many military-to-civilian books are written that way, but if you've taken the time to read this book or listen to it as an audiobook, you're most likely looking to understand the concepts more than the details. In today's world, reference material is much more appropriately provided online, where it can be updated more frequently and used when you are writing resumes or making applications.

Conclusion

This book is all about providing the precision guidance you need to understand the world in which you are about to engage. At the time of this writing, all the reference material you need can be found on our website, AbsolutelyAmerican.com, which will lead you to the detailed reference materials we provide through our Corporate Gray business unit.

At the same time, we have provided you with some specifics on how to handle certain common situations. We reserved the space in this book for specifics on what my experience has shown me are the most important things to execute. Opening the interview, closing it, and asking questions that show your interest in the what, how, and goals for the position you are seeking are the three most impactful things when it comes to the interview process and getting the best offer.

We also explained why you should discover, tier, and prioritize your core values and articulate the functional job duties you want to continue, add, and remove from your job descriptions. These concepts and techniques cannot be used effectively without an understanding of how to navigate career planning in the context of the rigid timelines that define the military-to-civilian transition process.

In the end, you alone are responsible for putting this information into action. My role in this is to help you understand it and assure you that my advice is sound and current. As I said previously, everything I have taught in this book has been evaluated against decades of experience in not only helping veterans succeed in getting their first jobs but doing so with organizations that have paid us a lot of money to bring veterans to them. I still work hands-on doing this work with our fellow veterans every day. So, if what I teach works when companies have to pay thousands of dollars on top of what it takes to hire you, how much more will it work when there is no third-party placement fee involved?

For my last thoughts, I encourage you to never be satisfied with the status quo. Everything in nature is either green and growing or

Conclusion

brown and dying. There is no status quo in nature, so if you are not intentionally moving yourself forward, nature is sliding you backward. This is consistent with the second law of thermodynamics, which states that any system that is not maintained will descend into chaos. Therefore, I encourage you to take time to make a habit of periodically examining where you are in each area of your life so you can adjust yourself appropriately to reach your objectives in balance.

Thank you for taking the time to read this book. I'm excited about the success that will come your way from what you apply to your life.

In the meantime, stay connected with us at AbsolutelyAmerican.com and CorporateGray.com and join our mailing list for the most up-to-date information available to help you grow and build a career that leads to long-term success and satisfaction. You can also follow me on LinkedIn and catch my Military Minute podcast at militaryminute.us, Apple podcasts, or wherever you access your favorite content.

THANK YOU FOR READING MY BOOK!

DOWNLOAD YOUR FREE GIFTS

Just to say thanks for buying and reading my book, I would like to give you a few free bonus gifts, no strings attached!

TO DOWNLOAD NOW, VISIT:

SCAN ME

I appreciate your interest in my book and value your feedback as it helps me improve future versions of this book. I would appreciate it if you could leave your invaluable review on Amazon.com with your feedback. Thank you!

www.ingramcontent.com/pod-product-compliance
Lightning Source LLC
Chambersburg PA
CBHW060547200326
41521CB00007B/517